CONNECTIONS

The EERI Oral History Series

Joseph P. Nicoletti

CONNECTIONS

The EERI Oral History Series

Joseph P. Nicoletti

Stanley Scott, Interviewer

Earthquake Engineering Research Institute

Editor: Gail Hynes Shea, Berkeley, California

Cover and book design: Laura H. Moger, Moorpark, California

Published by the Earthquake Engineering Research Institute

499 14th Street, Suite 320
Oakland, California 94612-1934
Tel: (510) 451-0905Fax: (510) 451-5411
Email: eeri@eeri.org
Website: http://www.eeri.org

EERI Publication Number: OHS-14
ISBN (pbk.): 1-932884-24-6
Library of Congress Cataloging-in-Publication Data
Nicoletti, Joseph P.
Joseph P. Nicoletti / Stanley Scott, interviewer.
p. cm. -- (Connections, the EERI oral history series ; 14)
Includes index.
ISBN 1-932884-24-6 (alk. paper)
1. Nicoletti, Joseph P.--Interviews. 2. Civil engineers--United States--Interviews. 3. Civil engineering--California--History--Sources. I. Scott, Stanley, 1921-2002. II. Title.
TA140.N53A5 2006
624.092--dc22

2006022188

Printed in the United States of America
1 2 3 4 5 6 7 8 10 2012 2011 2010 2009 2008 2007 2006

Acknowledgments

Funding for production of this volume was primarily provided by a generous donation from the Blume Foundation, established in honor of John A. Blume. Blume's oral history was published in 1994 as volume two in the *Connections* series. Additional support was provided by FEMA/U.S. Department of Homeland Security and the National Science Foundation.

Table of Contents

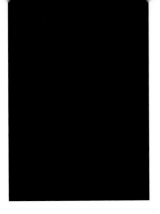

The EERI Oral History Series

This is the fourteenth volume in the Earthquake Engineering Research Institute's series, *Connections: The EERI Oral History Series.* EERI began this series to preserve the recollections of some of those who have had pioneering careers in the field of earthquake engineering. Significant, even revolutionary, changes have occurred in earthquake engineering since individuals first began thinking in modern, scientific ways about how to protect construction and society from earthquakes. The *Connections* series helps document this important history.

Connections is a vehicle for transmitting the fascinating accounts of individuals who were present at the beginning of important developments in the field, documenting sometimes little-known facts about this history, and recording their impressions, judgments, and experiences from a personal standpoint. These reminiscences are themselves a vital contribution to our understanding of where our current state of knowledge came from and how the overall goal of reducing earthquake losses has been advanced. The Earthquake Engineering Research Institute, founded in 1949 as a non-profit organization to provide an institutional base for the then-young field of earthquake engineering, is proud to help tell the story of the development of earthquake engineering through the *Connections* series. EERI has grown from a few dozen individuals in a field that lacked any significant research funding to an organization with nearly 3,000 members. It is still devoted to its original goal of investigating the effects of destructive earthquakes and publishing the results through its reconnaissance report series. EERI brings researchers and practitioners together to exchange information at its annual meetings and, via a now-extensive calendar of conferences and workshops, provides a forum through which individuals and organizations of various disciplinary backgrounds can work together for increased seismic safety.

The EERI oral history program was initiated by Stanley Scott (1921-2002). The first nine volumes were published during his lifetime, and manuscripts and interview transcripts he left to EERI are resulting in the publication of other volumes for which he is being posthumously credited. In addition, the Oral History Committee is including further interviewees within the program's scope, following the Committee's charge to include subjects who: 1.) have made an outstanding career-long contribution to earthquake engineering, 2.) have valuable first-person accounts to offer concerning the history of earthquake engineering, and 3.) whose backgrounds, considering the series as a whole, appropriately span the various disciplines that are included in the field of earthquake engineering.

Scott's work, which he began in 1984, summed to hundreds of hours of taped interview sessions and thousands of pages of transcripts. Were it not for him, valuable facts and recollections would already have been lost.

Scott was a research political scientist at the Institute of Governmental Studies at the University of California at Berkeley. He was active in developing seismic safety policy for many years, and was a member of the California Seismic Safety Commission from 1975 to 1993. Partly for that work, he received the Alfred E. Alquist Award from the Earthquake Safety Foundation in 1990.

Scott received assistance in formulating his oral history plans from Willa Baum, Director of the University of California at Berkeley Regional Oral History Office, a division of the Bancroft Library. An unfunded interview project on earthquake engineering and seismic safety was approved, and Scott was encouraged to proceed. Following his retirement from the University in 1989, Scott continued the oral history project. For a time, some expenses were paid from a small grant from the National Science Foundation, but Scott did most of the work pro bono. This work included not only the obvious effort of preparing for and conducting the interviews themselves, but also the more time-consuming task of transcribing, reviewing, and editing transcripts.

The *Connections* oral history series presents a selection of senior earthquake engineers who were present at the beginning of the modern era of earthquake engineering. The term "earthquake engineering" as used here has the same meaning as in the name of EERI—the broadly construed set of disciplines, including geosciences and social sciences as well as engineering itself, that together form a related body of knowledge and collection of individuals that revolve around the subject of earthquakes. The events described in these oral histories span many kinds of activities: research, design projects, public policy and broad social aspects, and education, as well as interesting personal aspects of the subjects' lives.

Published volumes in *Connections: The EERI Oral History Series*

Henry J. Degenkolb	1994
John A. Blume	1994
Michael V. Pregnoff and John E. Rinne	1996
George W. Housner	1997
William W. Moore	1998
Robert E. Wallace	1999
Nicholas F. Forell	2000
Henry J. Brunnier and Charles De Maria	2001
Egor P. Popov	2001
Clarence R. Allen	2002
Joseph Penzien	2004
Robert Park and Thomas Paulay	2006
Clarkson W. Pinkham	2006
Joseph P. Nicoletti	2006

EERI Oral History Committee

Foreword

This oral history volume is the completion of the interview sessions Stanley Scott (1921-2002) conducted with Joseph Nicoletti in the 1990s. Joseph Nicoletti and I updated and fleshed out that manuscript in interviews conducted in 2004 and 2005. Footnotes, tables, and photographs were added in this recent process.

Gail Shea, consulting editor to EERI, carefully reviewed the entire manuscript and prepared the index, as she has on previous *Connections* volumes, and Eloise Gilland, the Editorial and Publications Manager of EERI, also assisted in seeing this publication through to completion.

Robert Reitherman
Chair, EERI Oral History Committee
April 2006

Personal Introduction

My professional association and friendship with Joe Nicoletti began in 1958 when I joined the firm of John A. Blume and Associates, then located at 612 Howard Street in San Francisco. I had recently received my Master's degree from U.C. Berkeley, had worked for the State of California Bridge Department, and had returned from a post-graduate vacation. The job market was a little tight at that time and I had given up on finding a job with the private sector. Upon returning home from accepting a position with the State on a Friday afternoon, I had a message to call Joe Sexton. He told me there was an opening and asked I could start work on Monday. Joe Nicoletti was one of the four vice presidents of the Blume firm (the others being Don Teixeira, Joe Sexton, and Roland Sharpe), each taking on the role of managers for the design projects. My early work included being on the design teams for large projects such as the Ventura Marina, the San Francisco Federal Office building, and the College of San Mateo.

The first project I was given as my own was the structural design of an industrial research laboratory in Walnut Creek, California under the supervision of Joe Nicoletti. Although the primary building was only a single-story structure, it was an interesting project because of the combination of steel moment frames, reinforced concrete block walls, long span metal deck roof diaphragms and varying roof heights. The architect wanted a smooth-looking welded moment frame at a time that structural engineers were concerned about the quality of field welding. (Sound familiar? This was in the early 1960s.) The single metal deck did not work well as a diaphragm and required supplemental out-of-sight bracing. There was also a requirement to minimize potential concrete slab-on-grade cracking. I learned a lot working with Joe on this project, especially on how to work with architects. It amazed me how Joe was always up-to-date on my progress and was able to convey it to the client. There is an interesting postscript to this project. A year or so later, when taking my structural engineering exam, one of the questions was about designing long-span metal decks as diaphragms for a building remarkably similar to the Walnut Creek project. It occurred to me that Joe might have written that one.

After working with John Blume on projects in the last half of the 1960s, I returned to working with Joe during the post-1971 San Fernando earthquake period. This included structural evaluation of damaged instrumented buildings, evaluation of Veterans Administration hospital complexes using their newly developed criteria, earthquake vulnerability studies of the Puget Sound Naval Shipyard facilities and development of TriServices (Army, Navy, and Air Force) seismic design manuals for the Army. Joe was good at listening to my sometimes offbeat ideas. Our somewhat different backgrounds and personalities often led to interesting discussions that generally resulted in converging to a well developed consensus. For the Veterans Administration we used innovative methods of what now may be considered as performance-based engineering, greatly influenced by John Blume's concepts. For the Puget Sound Naval Shipyard project we developed a rapid dynamic analysis procedure, a forerunner of what became the capacity spectrum method, and which grew out of John Blume's reserve energy technique. In our work on developing dynamic analysis procedures for the TriServices manuals, we expanded the Puget Sound Naval Shipyard procedures into the CSM and introduced a method using inelastic demand ratios. I can clearly remember sitting across the desk from Joe, filling in the blanks for a preliminary inelastic demand ratio matrix for various structural elements and earthquake hazards in response a request from George Matsumura of the Corps of Engineers to develop a more user-friendly procedure for performance-based engineering. Joe and I seemed to complement each other in our structural engineering perspectives, and we worked well together. Even after I left the Blume firm, which by then had become part of URS, and joined Wiss, Janney Elstner Associates, Inc. in 1981, Joe and I remained friends and colleagues.

Recalling my association with Joe Nicoletti and the John A. Blume firm brings back memories of the Friday night end of the week get-togethers, John Blume's Fiscal Agent (a discounted bourbon), Blume-O-Grams, office picnics, and Saturday lunches at New Joes and Original Joes restaurants (no relation to Joe Nicoletti). I often think back to how fortunate it was to receive that call on the Friday afternoon in 1958 when the offer came in to join the Blume firm. My experience with the firm and people like John Blume and Joe Nicoletti had a profound effect on my development as a structural engineer and my participation in the field of earthquake engineering. Joe's acceptance of my engineering judgment gave me confidence in developing my skills in the field of earthquake engineering. I knew that I must be coming of age in my technical skills when Joe's primary critique of my reports and papers related solely to grammar.

It was a difficult decision to make when I decided to leave the firm in 1981 and join Wiss, Janney, Elstner Associates. The joys and independence of John A. Blume & Associates were slowly fading as the influence of the URS corporation style of management was taking over. When I left, the firm was still working on completing the TriServices manual on essential buildings (e.g., hospitals etc.). I continued to work with Joe to complete the manual through Wiss, Janney, Elstner. Our two firms then worked jointly on the manual for existing buildings and conducting workshop courses for the Army. After the Loma Prieta earthquake of 1989, we often ended up on opposite sides as consultants for litigation cases, so we had to keep at arm's length to avoid discussing our opinions. I missed not being able to debate Joe on the technical issues involved in these cases, as well as on other issues on the direction of our profession. A few weeks before writing this introduction I talked to Joe, and because these litigation cases have since been closed, we discussed some of the issues. I am looking forward to continuing some of these debates in the future.

Sigmund A. Freeman
September 2005

CONNECTIONS

The EERI Oral History Series

Joseph P. Nicoletti

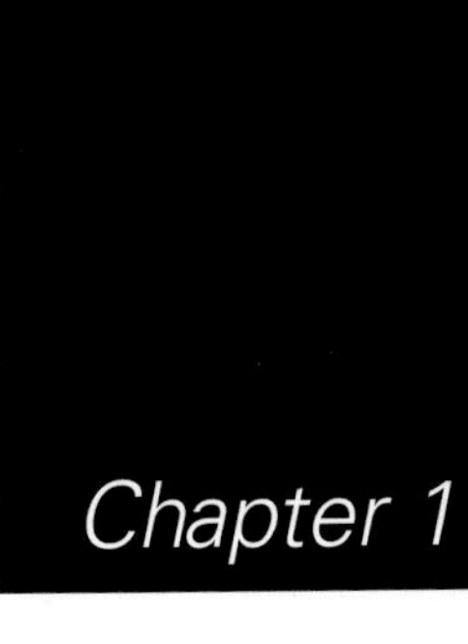

Early Years and Education

I have seen your College Board test results and you are going to be a civil engineer, and you are going to the University of California.

Nicoletti: Both of my parents were born in the small village of Ponte San Pietro on the outskirts of the city of Lucca in Italy. My father served his compulsory military training in Italy, with his class of 1906. Then in 1908 he was called back up for service because of the big earthquake in Messina, when they called on the Italian Army for disaster relief.[1]

My father came to the United States in 1911, and in 1918 he was drafted into the American Army, so in all, he served three involuntary tours of duty in the military service. While he was in the U.S. Army, he trained at Camp Fremont, which is now part of Stanford University. He took classes and applied for and received U.S. citizenship. He had been working for the Italian Swiss Colony Winery in Asti, California, but when he

1. The Messina-Reggio earthquake in Italy on December 28, 1908 and the resulting tidal wave killed over 100,000 people and wiped out the villages of Messina, Reggio Calabria, and dozens of other nearby towns.

came back from the Army, he no longer had a job, because in 1920 Prohibition had been voted in. While he was in the process of deciding what to do, a big earthquake occurred in northern Italy in 1920, right after he got out of the U.S. Army. It was a very severe earthquake in the area where his family lived. People were killed, and there was a lot of property damage.

My father went back to Italy to see his family. My mother and my sister were already there, and as a result of that visit, I was born in Ponte San Pietro in 1921. When I was born in Italy, both my mother and father were American citizens. My mother became a citizen when my father was naturalized. I had dual citizenship when I came to the U.S. in 1927, and I retained it until I renounced the Italian citizenship when I accepted a commission in the Navy in 1942.

Scott: I presume you were required to do that in order to take the Navy commission.

Nicoletti: Yes. All of my education was in the San Francisco Bay area. I attended grammar school in San Francisco, high school in Daly City, and junior college in San Mateo. Then I went to the University of California at Berkeley, where I graduated in 1943. When attending grammar school in San Francisco, we actually lived right across the county line to the south, in San Mateo County. Then during the Depression, I guess they weeded out the people who lived outside the school district, and I was one of them. So I had to leave Longfellow, the grammar school in San Francisco, after the fourth grade and attend Crocker grammar school in Daly City. That school was actually farther away from my home, which is why I was going to school in San Francisco. As a result, I graduated from Jefferson Union High School in Daly City, and then went on to San Mateo Junior College. After that, I went to the University of California.

Scott: Could you say something about your grammar school or high school years? Were there any special experiences, or special teachers, or early indications of your own interests and directions?

Nicoletti: I entered the first grade at Longfellow Grammar School in San Francisco. At the time I could not speak a word of English, but had a wonderful teacher, Miss Dwyer, whom everybody loved. She must have been near retirement age. She was very patient and supportive with me, and I could never forget her. At Jefferson High School in Daly City, the two teachers that made lasting impressions were Hugh Patterson and Dorothy Pendergast. Mr. Patterson taught geometry and insisted on formal Euclidian proofs for all of the problems. I had Mrs. Pendergast for four years of English, and I am eternally grateful to her for teaching me the fundamentals of English grammar and composition. In her fourth-year class, we had to read an English novel and present a written or oral report every week.

When I was in high school, I was not quite sure what I wanted to do. I had been in the Boy Scouts, and also in the Sea Scouts, and I had considered applying for one of the military academies. I found out, however, that it was quite difficult to get into West Point or Annapolis. To do that, you almost had to know a Congressman or a U.S. Senator who would recommend you. So instead, I was thinking about the Coast Guard Academy. I had read a lot of literature on the Coast Guard Academy, and it sounded quite interesting.

Before I went to Cal, I went to San Mateo Junior College, where my advisor was a professor by the name of Dan Reichel. He was a real father figure. Dan taught civil engineering and had graduated from the University of California in civil engineering as a classmate of Robert Gordon Sproul, who was Chancellor of the university while I was there.

When I first met Dan, he took me aside and said: "I have seen your College Board test results and you are going to be a civil engineer, and you are going to the University of California." And I did. Dan also introduced me to Don Teixeira, who turned out to be a very good friend of mine. We went through school, and then eventually ended up working for John Blume together. Don passed away about 1980. It was quite a loss.

Scott: Your advisor, Dan Reichel, obviously kept in touch with you.

Nicoletti: Yes. Dan Reichel and one or two other professors at San Mateo Junior College were quite influential in directing me in the way that they thought I should go. I think this was quite true in general of the staff at San Mateo Junior College. At that time, the college was quite a small school, so we had close contact with the teachers.

Sam Francis, another instructor at San Mateo Junior College, had the greatest impact on me. Sam was a math professor and a real teacher. I learned a lot from him in the math courses I took—calculus, advanced complex variables, differential equations, and so forth. He made math very interesting, very exciting. His son, also named Sam Francis, is a well-known modern artist. Another professor, Les Wilson, who came to San Mateo Junior College from the U.S. Geological Survey (USGS), was responsible for my continuing interest in geology and mineralogy.

Gold into Lead: An Early Glimpse of Atomic Research

Nicoletti: I recall that one of the assistant instructors in chemistry at San Mateo Junior College, Richard Weems, was also a graduate student working with Ernest O. Lawrence at the University of California, Berkeley. I remember his coming into the lab one day very excited, and said, "They just succeeded in converting gold into lead." At first I did not think that was so very exciting, but found that what they had actually done with Lawrence's accelerator at Berkeley was to break gold atoms down into lead atoms.

Scott: So it was not a joke, but an early successful experiment in atom-splitting?

Nicoletti: Yes. Then, when I went to Berkeley, Lawrence was still working in the chem lab with his little table-top cyclotron, and they had just started construction of the big cyclotron up on the hill. Later on, of course, the whole thing came back to me again when I was overseas during the war, and the first atomic bomb was dropped. I recollected back to that incident in my chem lab at San Mateo Junior College, when the instructor reported on the table-top cyclotron experiment. That had been in the early days of the work that led to the atom bomb.

Scott: About when did that happen?

Nicoletti: It was in 1940. Professor Lawrence went on to work with the atomic bomb project, and was instrumental in helping develop the bombs that were dropped on Japan during World War II.

The building of the big cyclotron up on the hill was based on the success of this little table-top one Lawrence had developed in the chem lab. I also remember that when the war started at the end of 1941, they immediately placed guards around the cyclotron. I had not realized the significance of that at the time, but I remembered it later on, when the atomic bomb was dropped. Then I could put all the pieces together.

Scott: They built a whole faculty around atomic expertise, and at first I guess pretty much around Lawrence. Then there were many others, such as Robert Oppenheimer, who I believe was instrumental in getting the Los Alamos, New Mexico site chosen and heading up those activities there. Fifty years later, Los Alamos is still run by the University of California. It seems like a strange anomaly, driving through Los Alamos on a remote site in the middle of New Mexico, and seeing the sign saying that the University of California runs it.

Nicoletti: I never met Robert Oppenheimer, but of course I have heard a lot about him. I did meet his brother, Frank, who was instrumental in setting up the Exploratorium in San Francisco. I was involved in several small projects done for him at the Exploratorium, and found him to be a very gracious and modest individual.

Engineering Education

Scott: Would you discuss the period when you were actively a student of engineering? Also, what are your recollections about the engineering program at Berkeley, where you earned your engineering degree?

Nicoletti: My engineering education started with the San Mateo Junior College days, when Dan Reichel and his classes in surveying steered me into civil engineering. When I went to Berkeley in the fall of 1941, however, I found it quite a change from the junior college.

Scott: Big classes, for example?

Nicoletti: Yes. The classes had sixty or seventy students, and you had very little contact with the teacher. Generally, I think that even though the professors at Berkeley were great engineers and quite well known in their field, they were not teachers in the same sense that the people at San Mateo were.

What impressed me the most was how impersonal the teaching was in Berkeley. I do think, however, that it instilled a sense of discipline. There was an awareness that the teachers did not care whether you passed the course or not. In fact, they would just as soon weed you out and make the class smaller. So you had to learn discipline.

Notable Professors at Berkeley

Charles Gilman Hyde

Nicoletti: One of the teachers I recall in particular was Charles Gilman Hyde, a professor of hydraulics and water supply and other related courses. His typical approach was to give out all of the class assignments at the beginning of the semester. Then he never talked about them again. His lectures were very interesting but had nothing to do with the subject matter that had been assigned.

It was up to you as a student to follow the study assignments and do the problems and turn them in on time. Otherwise you flunked. If you had any questions you went to the graduate student teaching assistant, you did not go to the professor, except as a last resort. Hyde was the extreme, of course, but most of the professors at that time worked pretty much on the principle that the student was given an assignment and it was up to the student to get that assignment done and done on time. I think that's the most important thing I learned from those days at Berkeley—discipline. I think it came in very handy in later years. Sometimes the best way to attack a problem is to get in at the bottom and do it.

Charles Gilman Hyde was a very impressive fellow. Nobody missed his lectures, although as I said, they never had anything to do with the assignment. He taught hydraulics and water supply, was well-traveled, and had slides of Roman aqueducts, and canals in Babylon, and so forth. His lectures were very interesting.

Scott: Did any other professors make an impression on you?

Charles Derleth

Nicoletti: Well, several professors stood out at Berkeley. One of them had to be Charles Derleth, who was the Dean of Engineering at that time. He was a very, very impressive professor. He was very small in stature, but had a reputation, and a great sense of humor. He taught several courses. Dean Derleth was quite well-known in his day as an educator as well as the foundation consultant on the Carquinez, the Golden Gate, and the San Francisco-Oakland bridges.

One incident I remember about Derleth also involved another professor, Francis Foote, who taught the survey classes. Professor Foote could get almost apoplectic when he was upset or excited. Once I had a roommate who spent more time worrying about his problems than doing them. We were both in Foote's class, and there were times when he had to copy my work in order to get his assignment turned in on time. Foote called us in one day and accused my roommate of copying my problems sets. My roommate admitted that he had. So Foote threw us both out of his class, and told us that we had to go see the dean to get reinstated.

At that time I had not met Dean Derleth at all. We had this appointment with him, and both of us were scared to death. When we got in to see the dean, he said to my roommate, "I understand that you have been copying Joe's problems sets?" Then the dean started to tell us how he carried half the football team at Columbia through school by letting them copy his problem sets and his assignments. He said, "You have to help each other," but he also said, "Do not be so obvious about it—Foote does not like it." So he sent us back and reinstated us in the class. After that we were less obvious about it.

Derleth made a big impression on me at that time. Later on I had him for a couple of classes, and he was delightful.

Bernard Etcheverry

Nicoletti: I was interested in structures. At that time the only minors available were things like transportation and surveying and irrigation. I selected irrigation because at that time irrigation offered more in structural design than the other options did. We had basic design courses in the civil engineering curriculum, and you could select a minor as a specialty. Most of the structural engineers from my time in Berkeley, the ones in practice today, came out of this irrigation minor.

Scott: I have heard that before from other engineers I have interviewed, but would you say a word or two more about why irrigation, at least irrigation as taught then, would give you so much structural design?

Nicoletti: Bernard Etcheverry was the professor of irrigation at that time. He now has a hall named after him, and was a wonderful fellow, already probably in his sixties at that time. He was a pioneer in western irrigation, and emphasized irrigation structures, dams, flumes, culverts, and settling basins. He gave us practical direction in the design of these structures. He had problem sets in which we actually had to design these structures, so we got good hands-on design guidance, which was missing in some of the other courses.

Howard Eberhart: Thesis Advisor

Nicoletti: Howard Eberhart is another professor who also stands out in my mind. He was the favorite of most of the young engineers because at that point he was one of the younger professors at Berkeley. His courses were in indeterminate structures. I never had any classes directly from him, although he taught portions of some of the classes I was taking.

At that time, Berkeley required an undergraduate thesis for the baccalaureate degree in civil engineering, and Eberhart was my thesis advisor, so I got to know him pretty well. I had spent the summer between my junior and senior year at Mare Island Naval Shipyard under training in shipbuilding, and I had become interested in launchings, so I did my thesis on the engineering aspects of ship launchings. And Eberhart was very interested, and in fact he gave me an A on the thesis.

I kept in touch with Howard Eberhart. He eventually retired from Berkeley, went over to Saudi Arabia, and set up an engineering school in the University in Riyadh. A 1944 accident requiring amputation of his left leg below the knee led him into research on human locomotion. Eberhart and his associates developed many of the principles used in designing artificial limbs and braces.

Finishing College During the War

Scott: You finished your engineering education after the U.S. entered World War II, didn't you?

Nicoletti: Yes. I mentioned that Don Teixeira and I graduated from San Mateo and went on to Berkeley together in the fall of 1941. I remember driving back from my parents' house on a Sunday, December 7, crossing the Bay Bridge and turning on the radio in the car. It was the news about Pearl Harbor. When war came, Don interrupted his classes at Berkeley to go into the Air Force, whereas I elected to go into the Navy program, called the Yarnell Program.

Shortly after Pearl Harbor, in December of 1941, the Navy realized that they would need to recruit college students with an engineering background to facilitate the shipbuilding and repair operations foreseen for the duration of the war. Naval ROTC graduates were slated for sea duty, and other engineering graduates were offered commissions in the Army Engineers or the Navy Seabees. Admiral Yarnell was responsible for the program that recruited college juniors and seniors in engineering for shipbuilding and repair duty in shipyards and advance bases overseas after they graduated.

Scott: So you were able to join the Navy and continue your education?

Nicoletti: Yes. The Navy offered provisional commissions and allowed us to finish our classes. When the war started, I had a year to go, so I received a commission in 1942, although I did not graduate until 1943. During the summer between 1942 and 1943, they sent me to Mare Island on temporary duty, where I received training in shipbuilding.

Mare Island Military Experience

Nicoletti: While I was at Mare Island that summer between my junior and senior years, I got a lot of the reference material for my thesis. Mare Island had some very successful experiences in ship launching, and also a few very unsuccessful ones. I was able to get a lot of good reference material there. I had access because I was in the Navy at that time, had spent several months at Mare Island, and had gotten to know some of the people there and actually participated in the launching crews of several ships.

Scott: Launching a large vessel is no doubt a very difficult maneuver. It means trying to shift something from one state to another very different state, in a tricky and fast-moving transition where things can go haywire in a hurry.

Nicoletti: Yes, launching is a dynamic process. The ships are generally built on sliding ways, and are in cradles. When a ship is ready to be launched, the cradle is put into contact with the building ways, but before that, the weight is carried directly to the ground by shoring without bearing on the sliding ways. Just before launching, the connection to the ground is removed, and then the weight of the ship is on the launching ways, which are greased and ready for the ship to be slid into the water. As the ship goes down the ways, the weight is pretty much equally distributed.

Then when a portion of the ship enters the water, but before there is appreciable buoyancy, the stern of the ship overhangs the launching ways, creating a "hogging" condition. As the ship moves further into the water, the stern will be supported by buoyancy, while the bow is still on the ways, creating a "sagging" condition. So you immediately have this second condition where the bending is in the opposite direction. Those conditions are quite critical and have to be studied in great detail. It is a dynamic problem, because the ship is moving and has inertia and velocity and so forth.

The other problem in launching is to absorb the kinetic energy that the ship has acquired moving down the ways. The ship has potential energy when it is on the ways, which is converted to kinetic energy as it slides down. That energy has to be absorbed somehow, you have to stop the ship once it gets into the water. The ship is not under power.

This was one of the problems they had at Mare Island with the battleship *California*. They had a device for slowing the ship down, which consisted of steel cable being run through hydraulic jacks so they could gradually slow it down and control the speed as it got into the channel. This had worked fine for smaller ships, but for the battleship *California*, they needed to multiply everything by a factor of three or four, to make the cable bigger and the force on the cables greater. The jacks became rifled with the lay of the cable as the ship was going down, and the cables unwound, snarled, and eventually broke.

The *California* went across the channel and as, they say, "half-way up Georgia Street." She got stuck on the mudflats on the other side, and it took every tug in the Bay to pull her loose at high tide. Mare Island was still living with that incident when I was there in 1942. The experience was almost as embarrassing as the nuclear submarine they sunk alongside the dock in the 1960s.

I should conclude this section on my formal education at San Mateo and at Berkeley by noting that we actually learned very little about earthquakes. Earthquake engineering was not even a term at that time, and I really did not learn much about earthquakes until I began working for John Blume.

John of course had been interested in earthquakes for many years. He had worked with Lydik Jacobsen at Stanford, and then had been with the Coast and Geodetic Survey. John had been interested in ground motion and its effect on buildings for a long time. Actually, John has been called the father of structural dynamics. One of his major interests was the dynamic response of structures to ground motion.

World War II

This Marine group flew every day and kept the Japanese pinned down on the surrounding islands.

Nicoletti: After I graduated, I was sent back to Mare Island Naval Shipyard, which was a very exciting place in 1942 and 1943. Frenzied repair work was being done on ships damaged at Pearl Harbor or in subsequent Naval battles. I was assigned to the Shipbuilding Department, where the new construction was not quite as frenzied as in the Repair Department. We were building submarine tenders and I eventually got to work on four of the five tenders that were built at Mare Island in the 1942-1944 period.

My immediate supervisor and mentor was Commander Jack Stewart who had graduated from Berkeley in naval architecture during the Depression and had spent some time with the Civil Conservation Corps and the Border Patrol in Calexico before the war gave him an opportunity to work in shipbuilding. Jack was very good with people, particularly the civilian supervisors at the shipyard, who were very proud of their skills and resented any interference from the Navy officers. Jack taught me a lot about friendly persuasion that helped me in dealing with people in the Navy as well as in my engineering career.

Duty at Mare Island was very interesting, and I eventually became the Hull Superintendent in the construction of submarine tenders. I got very interested in the ships, and decided that I would like to go out to sea on one of them. I had met the crew that had reported for commissioning the ship, and they also wanted me aboard, because I knew more about the ship than anybody in the crew.

So I wrote a letter, which the captain endorsed, and in two weeks I had orders to serve on the USS Howard W. Gilmore (AS-16), which we had built at Mare Island. In July of 1944, we left for the Pacific Theater. I spent eighteen months on the Gilmore, and in that time we were involved in setting up advanced submarine bases in the Marshall Islands and in the Philippines.

Scott: Tell a bit more about the kind of ship it was and what it did.

Nicoletti: The *Gilmore* was a submarine tender. We were the flagship for a submarine squadron, which consisted of eighteen submarines, a rescue vessel, and a floating dry dock. We formed a task group that would set up a base in an advance area in order to cut down the patrol time for our submarines. During the early part of the war, when we did not have these tenders or the advance bases, the submarines had to spend most of their time cruising from say Pearl Harbor to a patrol area and then back. So they had very little time to spend in the patrol area.

The idea was to extend that time by setting up these tenders as advance bases as close as possible to the war zone. When we were in Majuro in the Marshall Islands as a matter of fact, we were surrounded by Japanese-held islands—Wotje, Jaluit, and Maleaolap. They were kept in check by a Marine air group, which was also based on the island of Majuro, where we operated. This Marine group flew every day and kept the Japanese pinned down on the surrounding islands. Every once in a while, of course, a Marine plane would be damaged or shot down, and we would have to go out and rescue the pilot.

Scott: How long were you out there?

Nicoletti: We were overseas for fourteen months. Most of the officers and the senior petty officers were qualified for submarine duty and many had been with the old S-boats in the pre-war Asiatic Fleet in China and had a lot of interesting stories to tell. I was a deck division officer and considered as "ship's company," as opposed to the repair divisions that serviced and repaired the submarines. However, because of my prior shipyard experience, I was occasionally "loaned" to the repair group to help out, particularly when they had submarines with structural damage. An interesting sidelight of working on the submarines was that we had to accompany the crew on the post-refit trial to get their approval on the work that we had done. We were issued special orders and received submarine pay and allowances for the two or three days of trial run.

Because the quietness of the submarine had to be verified after maintenance and repairs, the submarine would be taken out from the harbor and set down on the bottom about 100 feet deep. Everything would be switched off, lights and everything. Then one by one, the systems would be turned on while a ship overhead was monitoring to detect any telltale sounds.

One time we took a submarine out for a test after a hatch damaged by a depth charge had been repaired. I don't think the captain wanted to take his vessel back into action that soon, because when he took it to 200 feet, the usual maximum operating depth, and it didn't leak, he took it to 250 feet and then 300 feet, the maximum depth for structural safety. There was a little trickle of water—he finally got it to leak. At that pressure, the steel groans and creaks, an eerie sound.

End of World War II

Nicoletti: I spent eighteen months on the *Gilmore*. We had been in the South Pacific for fourteen months when the war ended. We came back and were attached to the Atlantic Fleet. Before I got out of the service, I spent the last three or four months with the Atlantic Fleet in New York, on the same ship, the *Gilmore*. In the Pacific, we had first been attached to the Fifth Fleet, then to the Seventh Fleet, and then eventually to the First Fleet out in the Atlantic.

Scott: So the ship actually went through the Panama Canal, or somehow got to the Atlantic area?

Nicoletti: Well, the ship did go through the Panama Canal. But before I left Mare Island, I met my wife, Josephine, who was a Navy nurse there, and became engaged to her before I left. So when the ship came back from the Pacific, I took two weeks leave in San Diego to get married. My fiancé had been transferred to the Naval Hospital at Farragut, Idaho, so we were married in Coeur D'Alene. Then I rejoined the ship in New York. For many years I complained about the fact that I had been deprived of the opportunity of seeing the Panama Canal. Finally, in 1998, my wife and I went through the Canal on a cruise ship. The canal is an impressive civil engineering monument and certainly worth seeing.

Scott: You went to New York and you were with the same group there?

Nicoletti: Yes. Our whole squadron, with all of our submarines, was shipped to New York.

Scott: That was just for a relatively short period of time?

Nicoletti: The war was over, and they were reorganizing the different fleets. The ship I had been on eventually ended up down in Key West, Florida, but that was after I left.

USS *Oglethorpe*

Nicoletti: I left the *Gilmore* in early 1946, and was assigned briefly to an attack cargo ship, the USS *Oglethorpe* (AKA-100). The ship was part of an operation called the "Magic Carpet." The Magic Carpet was intended to bring back overseas personnel and supplies that had been in the war zone.

The *Oglethorpe* was an attack cargo ship equipped with about twenty-four landing barges that could be used to transfer cargo and equipment to the beachhead during an invasion. When I was attached to her, she had already made one or two trips. I joined her in Seattle, and we were all set to go out in the Pacific again, when our orders were changed and we were sent to San Diego to train Marines in amphibious landings. I was the First Lieutenant on the *Oglethorpe*. The First Lieutenant is the third in command on a ship, and I had four deck divisions, with about twenty-five or thirty people to a division, so I had about 100 men. The best thing about this assignment was that, as a department head, I was the duty officer every four days, but I did not have to stand any watches and that was a welcome relief.

We had twenty-four landing craft, and the idea was to train Marines in landing operations down in San Diego. We did that for a couple of months, and then our orders were changed again and we were sent to the Oakland Naval Supply Depot. The *Oglethorpe* was loading up cargo to go out to Bikini Atoll for the atomic bomb test. By that time, I had earned enough points to retire from active duty, and so I did. My replacement showed up the day before the ship sailed. If he had not shown up, I would have had to go out with the ship.

Scott: When did your retirement from active duty take place?

Nicoletti: In June of 1946 I was separated from active duty. I maintained an inactive status in the reserve. They kept asking me to join an active unit, which I declined, but I did maintain an inactive status until I had about twenty-five years in the reserve. Then they told me that I either had to resign my commission; retire without any benefits; or join an active unit. I elected to retire without benefits.

I am in the retired reserve without any benefits. My reserve status has come in handy, because a lot of my early work with John Blume was with Navy installations—bases here in the U.S. and in places like Guam and the Philippines—and the Naval identification card was very handy in getting me into places that were more difficult to get into if you did not have any such ID.

Chapter 3

With the Blume Firm

I set out with the San Francisco telephone yellow pages, started visiting all of the listed construction companies, and had worked my way to the C's without any luck.

Scott: After you retired from active military status in June 1946, I assume you looked around for employment. There were lots of other people were looking for jobs then, too, weren't there?

Nicoletti: Yes, I found that there were many ex-service personnel looking for work. So I first checked with Bert Summers, who had given me a summer job five years before. It was back in the summer of 1941, after I had graduated from San Mateo Junior College and decided to find some temporary work in construction. I set out with the San Francisco telephone yellow pages, started visiting all of the listed construction companies, and had worked my way to the Cs without any luck.

When I came to Clinton Construction, they told me that they did not have anything for me, but that their former chief estimator, Bert Summers, had formed a small construction company, Nielsen, Erbentraut, and Summers, and might be looking for a field engineer. In short, back in 1941 Bert Summers, who was a civil engineering graduate from MIT, gave me my first engineering job. The company was constructing facilities for newly inducted draftees at Forts Cronkhite, Barry, and Baker in Marin County. Bert was a member of the Structural Engineers Association of Northern California and, until his death in 1999, every time we met at one of the meetings, he would remind me of my first job.

Anyway, when I found that jobs were scarce in the summer of 1946, I called on Erbentraut and Summers (Nielsen had subsequently passed away), but at the time they had no work for me. So I accepted a job as a steel detailer for electrical transmission towers with Bethlehem Steel in South San Francisco. The work was demanding and challenging (the discipline learned at Berkeley came in handy). After a month or so, however, a call came from Erbentraut and Summers. It was a welcome reprieve.

Scott: Steel detailing was good experience, but you wanted to get into construction?

Nicoletti: Yes. Erbentraut and Summers were building a wind tunnel at Moffett Field, and I was offered a job as a field engineer. I worked down at Moffett Field for almost a year and finished the wind tunnel. Then the next project I was going to be assigned to was out in Hawthorne, Nevada, and since our first child was only a few months old, I decided I did not really like that.

Starting with Blume in 1947

Nicoletti: In the meantime I had been in touch with Don Teixeira, whom I knew from our days at San Mateo Junior College. Don was working for John Blume, and told me that John was looking for someone. So I came up to San Francisco and met John. John had started his own company and was building up a staff. He offered me a job and I accepted—I think I was the fifth employee that he had hired by that time. When I retired in 1987, after forty years, I had held just about every position with John Blume and the successor company, from junior engineer to president. I was president when I retired.

Scott: Your association with Blume started in 1947, and lasted over fifty years?

Nicoletti: Yes, it began in February 1947, on Valentine's Day. That's an easy day to remember! John Blume had started the firm in 1945—it was called John Blume, Structural Engineer. He only had one or two people, at the outset, Don Teixeira being one of his first employees.

At the time I joined the firm, John had quite a bit of work. He had previously worked for Standard Oil, and they, with a couple of other oil companies, were forming the Arabian-American Oil Company, ARAMCO. After the war, there was a big construction program during which they were building up facilities in Saudi Arabia. The chief engineer for ARAMCO was a man that John had worked with in San Francisco for Standard Oil. So he came to John to help him design the facilities. Over a period of about five or six years, we did a tremendous amount of work in Saudi Arabia. Then around 1952, we started doing work with the Navy when they were building up their facilities in the San Francisco Bay area.

Scott: Maybe you could take up those various other projects a little later, but first say more about the organization of the Blume firm, and its various changes over time. You joined the Blume firm in 1947, and I believe you have been an integral part of the firm, or closely associated with it, ever since. Would you briefly trace the Blume firm's organizational history, and your relationship with the firm in its various forms? According to Blume's oral history[2] he was sole owner until the firm's incorporation in 1957.

Nicoletti: Yes. John decided to form a corporation in 1957, to formalize what he was then calling John Blume and Associates. Several of us were considered to be associates of the firm, but in 1957 we incorporated to form John Blume and Associates, Engineers, and John made four of us vice-presidents—myself, Don Teixeira, Roland Sharpe, and Joe Sexton. And John was president, of course.

In 1971, we decided to merge with the URS Corporation. We traded all our stock for URS stock. We then became URS/John A. Blume and Associates, but we were able to operate in pretty much the same way we did before.

Decision to Merge

Scott: Say a little about the Blume firm's decision to merge with the URS Corporation. What led up to that, and why was that done?

Nicoletti: Well, one of the problems we had, starting with the initial incorporation in 1957 as John Blume & Associates, Engineers, was related to our ownership of stock in the company. At that time, of course, it was closely held stock—in other words, the stock could only be held by members of the firm. In order for closely held stock to have any value, you have to have a market for it, particularly if somebody wants to retire, or somebody dies, or wants to leave the firm for any reason. The person leaving has to have a market for his stock, and it has to be an established market at an established price. The way this is done with closely held companies is by a stock purchase agreement. That is an agreement which everybody agrees to when they acquire the stock. They agree that they will purchase somebody else's stock if he leaves or retires.

Our stock ownership was quite unbalanced, because John had most of it, and the four of us only had a small percentage. So it was very awkward, almost impossible for us to rationalize how we could buy John's stock, if John should decide to retire, or if he should die. We tried for several years to achieve a stock purchase agreement and we were unsuccessful. We could never agree on the agreement.

So the URS offer was attractive because we could then trade our stock for URS stock, which was on the market, it was public stock. So this was a way out of this dilemma. We looked into it, and several other engineering companies of good reputation had joined URS, so it looked like a good way to end our dilemma. That was the principal reason we did it.

Quest for Profitability

Scott: Was your management still pretty independent after the merger? Did you as the Blume organization continue to operate largely on your own?

Nicoletti: Well, when we joined URS we were promised autonomy, and for a while we had it.

2. *Connections: The EERI Oral History Series: John Blume*, Stanley Scott Interviewer. Earthquake Engineering Research Institute, 1994.

Scott: How long did you continue to operate autonomously?

Nicoletti: Almost ten years, from 1971 to about 1980. But then URS gradually started to interfere more and more, and actually to take away some of the benefits and began to direct our activities. They were trying to expand our activity. "You know, if you can make a certain amount of money doing this, maybe you could make twice as much by expanding into another area." I think URS eventually expanded into areas where they did not have the expertise, and I think that this was their downfall. Plus the fact that they also did some creative bookkeeping that got them into trouble.

This effort was particularly directed toward trying to make the operation as profitable as possible. Starting about 1980, there was a lot of emphasis on making the stock more attractive and increasing earnings. Eventually, however, this hurt the company. It led to the company's decline, and replacement of the URS management. As a matter of fact, what they had been doing did not meet with the approval of the SEC, the Securities and Exchange Commission. The URS management was actually penalized. They were forced to resign and are prohibited from engaging in new activity of this type. Now, I think the company is in a rebuilding mode.

Scott: The activity that led to the difficulties was related to trying to make the stock more valuable?

Nicoletti: Well, I think the URS management was then composed primarily of stockbrokers rather than engineers. URS started to go into an expansion mode, buying up other companies, and getting into other fields. For example, they bought Evelyn Wood, which is a speed-reading organization, and they bought an interest in a video training company. So they were going into other fields where they thought there was an opportunity to enhance the value of a company and then eventually sell it. Some of the ventures worked out and some of them did not. Eventually, more of them did not work out than did, so they ended up with a large deficit.

Retired, But Still Active

Scott: Did their ventures and the deficit affect your part of the operation?

Nicoletti: Yes, they started getting more and more restrictive on what we could do and could not do. That was also the primary reason why I decided to retire at the end of 1987. About that time, URS got into trouble with the SEC and with some of the other stockholders, and so the change in management took place.

Since that time, the company has been restructured and they have been rebuilding the reputation they had before. It is interesting that even before URS ran into trouble, several of us in the Blume group got together and decided how we would like the leadership of our group to be passed on. Actually, Marty Czarnecki, the man that we had picked for the job back in 1980, headed up the San Francisco office until URS merged us with Dames & Moore in 2001.

Scott: Say a little more about your own retired-but-still-active status.

Nicoletti: When I retired at the end of 1987, I had intended to be an independent consultant, and for a few years, I did operate in that way. I did some consulting for other engineering firms and I maintained my personal relations with several organizations that are better known by their acronyms but I'll name them in full: ATC, the Applied Technology Council; BSSC, Building Seismic Safety Council; Caltrans, California's state transportation department; EERI, Earthquake Engineering Research Institute; SEAOC, Structural Engineers Association of California; ASCE, American Society of Civil Engineers; and BCDC, the Bay Conservation and Development Commission (for San Francisco Bay). However, I was spending most of my time on URS projects and I was still using my old office pretty much on a regular basis. In the years following the Loma Prieta earthquake in 1989, I became even more involved in seismic evaluation and rehabilitation projects with URS.

The company's legal department became concerned about potential problems with the IRS regarding my consultant status, so in 1993 I was reinstated as a part-time employee. This gave me freedom regarding my office hours and also made me eligible for vacations, sick leave, and other company benefits, while it allowed me to continue my personal association with the other organizations.

Chapter 4

Major Projects Prior to the San Fernando Earthquake

Over the next twenty years or so, we were involved in the seismic analysis and design of more than seventy nuclear power plants.

Nicoletti: I have not discussed the Blume firm's major projects, and can do that now.

Scott: Yes, that would be a good idea. It would help if you could take them up more or less in chronological order, starting with the earliest you worked on.

ARAMCO Work in Saudi Arabia

Nicoletti: I have already mentioned that my years with the Blume firm started with an ARAMCO project, and that ARAMCO was our biggest client. Over a period of about five or six years, we did a tremendous amount of work in Saudi Arabia. We did waterfront projects, we designed oil loading piers and wharves. We did all sorts of buildings. We did all of the original buildings in Dhahran and Dammam, and in Ras Tanura and Abquaiq.

We were part of a design group, and all of us worked in the same office. There was an architectural firm, a mechanical and electrical engineering firm, and ourselves doing the civil and structural engineering. We did all of the work for ARAMCO during that period, until ARAMCO moved their offices out of San Francisco—first to New York, and then eventually to Dhahran, in Saudi Arabia.

In 1977, when we were applying for some work with the Saudi government, I finally had an opportunity to go to Saudi Arabia. For the first time, I saw the facilities that I had designed back in the late 1940s and early 1950s. My very first assignment with John Blume had been to design a small boat pier in Dammam.

Telephone Company Buildings

Nicoletti: Also during that period—in the late 1940s and early 1950s—we did quite a bit of work for the telephone company, for Pacific Telephone and Telegraph. They were converting to the dial system, so they had to build new facilities all over California; "Community dial buildings," they were called at that time. Some of them were quite small, about the size of a one-room office, and some of them were quite large.

Over a period of five or six years, we did more than two hundred of these buildings all over California. The average size was probably about 10,000 or 15,000 square feet. The largest one we did was a ten-story building in Modesto. These buildings were quite simple, and we were able to turn those jobs out pretty efficiently. During this period we also designed the microwave stations for the telephone company. These were buildings and towers situated on mountain tops, for microwave links stretching from San Luis Obispo to Santa Rosa, and from San Francisco to Salt Lake City. I designed the installation on top of Mt. Rose in Nevada at an elevation of 10,000 feet. It is designed for wind loads of 200 miles per hour.

U.S. Navy and U.S. Army Corps of Engineers

Nicoletti: After ARAMCO and the telephone company, our next major client in the early 1950s was the Navy. We managed to get a lot of work at Mare Island, where I had been before, and of course was familiar with the personnel and the facility. So I ended up being the project manager for almost all of the Navy work we did. We also eventually did work at Hunters Point and Port Chicago in the San Francisco Bay area, and later in Hawaii, Guam, and the Philippine Islands.

About 1957, the Navy built a new air base at Lemoore, California, which is down near Fresno in the San Joaquin valley. The Navy put together several design teams, and we were part of the team that did the Operations Area Facilities. Over a period of a few years, we designed six hangers, two overpasses for aircraft to go over the roadways, an operations building, a control tower, and numerous other smaller buildings.

We also did quite a bit of work for the U.S. Army Corps of Engineers. At that time, the military works branch of the Corps of Engineers was in San Francisco, but they are now in Sacramento. We did quite a bit of work for them at the Benicia Arsenal, and Hamilton Air Force Base, and later on at Travis and Castle Air Force Bases, all in California.

Scott: That was in the 1950s?

Nicoletti: Yes, and the work extended on into the early 1960s.

Atomic Energy

Scott: In his EERI oral history, John Blume talked a good deal about various projects for the Atomic Energy Commission.

Nicoletti: In about 1964, I believe we had two major projects, one involving the Atomic Energy Commission's Nevada Operations Facilities and the testing of nuclear devices at the Nevada test site near Las Vegas, and the other was the design of the Stanford Linear Accelerator for Stanford University in Palo Alto.

Our firm was initially selected to help the AEC establish a "threshold of damage" for buildings in Las Vegas and some of the towns near the test site. An underground nuclear test generates energy something similar to an earthquake, so we were retained to assess whether buildings in Las Vegas could be damaged. We maintained an office with about fifteen people in Las Vegas for almost thirty years, until 1994. Then the project was shut down because underground testing had been virtually eliminated. Our contract was phased out in the fall of 1994, and we were given about six months notice. We were able to help most of our employees relocate to other employment in the Las Vegas area.

During the earlier years our problem was to try to establish the threshold of damage, so we monitored important buildings in Las Vegas and Tonopah and other close-by areas. We would estimate the damage threshold of these buildings, instrument them to record the response to the ground motion, and then compare the estimated response with the instrumented and recorded response. The Atomic Energy Commission was gradually increasing the size of the devices that they were testing, until we established a threshold of damage, the point where it was actually damaging the buildings. The intent, of course, was to determine that point. After that, we were primarily involved in monitoring new construction and making sure that the ground motion did not exceed the threshold at which damage would occur.

Scott: New construction within a certain radius of the test site?

Nicoletti: In a radius of about 120 miles of the test site. Subsequently, we were also given the instrumentation responsibility, which had originally been handled by the U.S. Geological Survey (USGS). We eventually were doing all of the instrumentation—both the free-field instrumentation and the building instrumentation.

Scott: How do you monitor new construction? Most construction does not get that kind of scrutiny.

Nicoletti: No, it doesn't, and the monitoring was actually not publicized. We did it almost without recognition. Of course, the building department was very grateful for this, because we were actually helping the building departments in these towns. We established a threshold, or the size of the device that could be tested and thus the shaking it would generate without exceeding the threshold of damage to the existing buildings. So we knew that if the buildings complied with the current building codes or better, they would not be damaged.

Our function was primarily to look for unusual buildings or unusual features that might respond in a way that would make the buildings more vulnerable to damage from the nuclear test shots—more vulnerable than ordinary common construction, or at least construction built only to the code. Of course, many of the new Las Vegas hotels then were unusual in some respects. So we looked at the new construction and advised the building department on changes that should be recommended to the designer. In most cases, these were in fact recommended and implemented. That was our primary work in Las Vegas.

In the early days, when we were establishing the threshold of damage, we had damage claims come in that we would have to run down and evaluate. There was not very much of that after the first few years, because we had pretty much weeded out the false claims.

Scott: Can you say a little more about the ground motion that resulted from these underground shots?

Nicoletti: Yes. It is surprising how much motion these underground shots cause—they are very much like small earthquakes. The signature is a little bit different from that of a natural earthquake, but the response of buildings is quite similar. During underground nuclear testing, we have experienced water actually being splashed out of the swimming pools located on the roofs of some of the hotels in Las Vegas. So the motion is quite similar. It was always very interesting to watch the gamblers. They don't react to the earthquakes at all. They just look up from their slot machines and then go right back.

Stanford Linear Accelerator

Nicoletti: The other major project that we started at about the same time was the Stanford Linear Accelerator. Started about 1965, it is a two-mile tunnel with some unusual features, designed and built as a cut-and-cover structure.

Scott: Was its proximity to the San Andreas fault one of the unusual features? It probably had quite a few other unusual features, too.

Nicoletti: Yes, it is very close to the San Andreas fault. Also, it was also constructed without any expansion joints.

Scott: That is remarkable—a two-mile concrete tunnel with no expansion joints!

Nicoletti: Yes, it is, 10,000 feet of reinforced concrete without an expansion joint. So the concrete technology was very critical. We recommended a granite aggregate, and were very careful about how the concrete was poured. Then we grouted whatever shrinkage cracks showed up in the curing of the concrete. Then of course we covered it with earth, and that pretty much stabilized it all. I don't believe that they have had any problems with leakage or any subsequent cracking. It was done by sequenced pours. In other words, you pour a section, then skip a section, which they would come back later and pour. Then they grouted the joints.

Another critical feature of the accelerator was its need to be perfectly straight. I think the tolerance is something like 2 millimeters in 10,000 feet, vertical tolerance or horizontal tolerance.

Scott: That sounds exceedingly demanding. How was it done?

Nicoletti: The first step was trying to establish a fairly stable structure. Most underground structures are stable. And then it is done by adjustment within the supports for the electron beam guides. The beam is very small, and the guides have to be adjusted within a tolerance of about 2 millimeters in 10,000 feet.

It involved some interesting designs, and also some interesting concepts. One feature was the so-called "beam switchyard." The purpose of the accelerator was to start electrons going down this 10,000-foot passage. Along the way, they were periodically accelerated. These electrons were used for research in various experiments. So at the end of the accelerator, when the electrons had achieved their terminal velocity, there was an electronic switching device that allowed them to be diverted to various buildings where the research was being conducted. The electrons might be diverted to one building, and then the next experiment might be over in another building, and so forth.

Scott: The various experiments would have targets set up in the buildings.

Nicoletti: They had all sorts of experiments. They would run the electrons through water or through different devices. We had to design what was called an "end dump station." This was so that if the accelerated electrons were not going to an actual experiment, but they didn't want to shut the accelerator down, there was something that would absorb them temporarily while another target was being set up. This was a massive structure, and involved not only earth cover, but also heavy steel to absorb the energy of the electrons. As a matter of fact, I think they used great big blocks of steel that they had gotten from a steel mill. You needed the mass to absorb the energy of the electrons. So that was another very interesting feature.

I had very little to do with the accelerator myself. As a matter of fact, Roland Sharpe was sent down to be in charge of the design, so I had to take over some of his duties here in San Francisco. Then when Rol came back we reorganized the firm somewhat. He became the executive vice-president, and I was the senior vice-president and chief engineer, so we had some changes in status.

Scott: So while the accelerator project was active, it was big enough and lasted long enough that I guess you set up an on-site office with Roland Sharpe in charge.

Nicoletti: Yes. For several years we had about fifteen or twenty people in our design group down there. The work was done in a joint venture called ABA. Guy F. Atkinson was in charge of the construction management, and AeroJet General, which was part of AETRON, was in charge of the instrumentation. They designed the instrumentation, working with the Stanford scientists. We did the civil and structural work. We also had an architect, Charles Luckman, quite a well-known architect from Los Angeles. He was not part of the joint venture, but we hired him as a subcontractor.

Nuclear Power Plant Projects

Nicoletti: I would like to say a few words about the firm's involvement with the design of nuclear power plants and other nuclear facilities. Our first exposure to nuclear power plants was shortly after 1960 when General Electric asked us to provide dynamic seismic analyses for two nuclear power plants in Japan, Tsuruga and Fukushima. This led to similar assignments with General Electric, Westinghouse, and other nuclear contractors for plants throughout the U.S. and Europe.

Scott: Your initial experience with these plants was in Japan, and later you were involved with similar plants in the U.S.?

Nicoletti: Over the next twenty years or so, we were involved in the seismic analysis and design of more than seventy nuclear power plants. John Blume was appointed to the first nuclear plant review board by the Atomic Energy Commission, but we were not precluded from providing consulting services on any project as long as John did not participate in its review. Rol Sharpe assisted John in these reviews and Joe Sexton was the project manager for the nuclear power plant work, while Don Teixeira and I took care of the conventional design and consultation projects. When Joe Sexton left the firm in 1971, as senior vice-president and chief engineer, I inherited the nuclear work. Fortunately, I had very good people in place to help me: Ron Gallagher for the nuclear power plant work; Dilip Jhaveri for nuclear waste analyses we were doing at Savannah River and Hanford; and Don Teixeira and Bob Van Blaricom for the conventional design work.

Scott: I seem to remember that your firm had something to do with the Diablo Canyon Nuclear Power Plant for PG&E?

Nicoletti: Yes. In the fall of 1967, we were asked by Dick Bettinger, chief civil engineer for Pacific Gas and Electric, to help them with structural design of the two turbine buildings for their proposed nuclear power plant at Diablo Canyon near San Luis Obispo, California. He said they had wanted to do the entire design in-house, but they had decided to farm out the design of the turbine buildings since they were to be conventional structures with no nuclear considerations.

We had almost completed the design in accordance with the UBC, the Uniform Building Code, when PG&E informed us that they were adding provisions for nuclear fuel storage at one end of the buildings. They thought that special provisions could be made locally for the storage area, without affecting the remainder of the turbine buildings. At this point, the Atomic Energy Commission had been replaced by the Nuclear Regulatory Commission (NRC) as part of the Department of Energy, and the NRC ruled that the entirety of both turbine buildings must be designed to nuclear regulations. The seismicity study prepared for PG&E had established design ground motion of 0.40g, based primarily on data available for the San Andreas fault. We proceeded to redesign the turbine buildings, and we also assisted PG&E in the design of the reactor buildings and other miscellaneous structures.

When the project was again about ninety percent complete, a seismologist studying drill cores from offshore oil exploration in the area detected what became known as the Hosgri fault, several miles offshore and parallel to the more distant San Andreas. Re-evaluation of the site seismicity increased the design ground motion to 0.70g. At this point, with design and construction costs escalating rapidly, PG&E decided to bring in the Bechtel Corporation to expedite the design and approval process. Whereas PG&E and Blume had previously debated the multitude of changes required by the NRC, now Bechtel's instructions were to do nothing that would slow the completion and approval of the project. Consequently, in the name of safety, many things that bordered on the ridiculous were done at the request of the NRC.

For example, in the turbine buildings, with relatively flexible diaphragms, support response spectra were required for the design of all piping supports. This required time history analyses of structural models with thousands of nodes and the extraction of response time histories at many of the nodes. The models were so large that they could be accommodated only at the largest CRAY computer, and our computer service billing reached as high as $250,000 per month. I believe that PG&E's initial budget was about $500 million, but it took ten years and about $5 billion to complete the project. In spite of the adverse publicity associated with nuclear power, I believe that Diablo Canyon has operated very successfully since 1986 and returned much of its construction cost.

Wells Fargo Building in San Francisco

Nicoletti: About that time, we were also approached by John Graham, an architect in Seattle, who had a commission to design a forty-three-story building in San Francisco. At that time, it would be the tallest building west of Chicago, and he had some reservations about it. He had his own engineering department, but they had no experience with the design of tall buildings, so they wanted us to help them with the earthquake design of this building. It is now called the Wells Fargo Building, at Market and New Montgomery. We did the analyses and the frame design. We found a graduate student at the University of California, a fellow by the name of Ian King, who was working on a computer program to handle tall buildings, called HIGHRISE. We funded his work on this program, which was developed for this particular building and used in the design.

Skidmore, Owings and Merrill were designing some highrise buildings in San Francisco during this period. They were working with someone else at Berkeley and developed two other programs, FRMSTC and FRMDYN. These are two-dimensional programs for static and dynamic analysis, and they became available about the time that we finished our design. So we used those to confirm our design. The HIGHRISE program was a tri-dimensional program, but it ignored some of the secondary response effects, such as the elastic shortening of the columns due to axial loads. It had more capabilities than these two-dimensional programs, but had some limitations. So the Wells Fargo Building was probably the first of the modern buildings designed with computer seismic analysis and built in this area, or perhaps on the Pacific Coast.

It had some interesting features that later on were copied in other buildings. First of all were the computer analyses, then also it was designed to be entirely field-welded. In other words, the steel moment frames were field-welded to make the necessary rotation-resisting (moment-resisting) beam-column joints, whereas everything up until then had been field-bolted or field-riveted. It had box columns, built-up columns of steel plates welded into square cross sections, because the available single-piece rolled shapes had large capacity for a moment in one direction, but very little in the other direction. So these box columns were ideal for corner columns where you had large moment demands in two orthogonal directions.

Ultrasonic Weld Inspection

Nicoletti: It was also the first time that ultrasonic inspection was used to make field welding much more reliable. Prior to the development of ultrasonic inspection, nondestructive inspection had to be done by X-rays or magnetic particles. These methods are very slow and difficult. The X-ray method in particular was a pain in the neck for field welding because you had to get the people out of the area and then you obtained these X-ray images, which had to be sent somewhere to be developed, and when you got them back you could not tell what you were looking at.

Scott: But the ultrasonic inspection actually worked well?

Nicoletti: It was quick and it worked well. You could calibrate it and then tell immediately where any defect was. So you would have the welder remove the defect and re-weld it, and have it reinspected.

Scott: I guess testing is critical in welding technology. Welding can be very good, but each weld needs to be checked to be sure that is done well?

Nicoletti: Right. Ultrasonic was the big jump in the ability to do that. We were approached by people who were promoting it, and we were sold on it. It was a big breakthrough. I think this was the first major project to use ultrasonic testing for field welding. Shortly thereafter, field-bolting almost disappeared, and welding became the way to go. In any event, word of our design and construction practices on the Wells Fargo Building reached Los Angeles, and before our design was complete, we were approached by Ed Teal, then Chief Structural Engineer for A.C. Martin and Associates. They were starting design on the forty-story Union Bank Building and wanted us to provide consulting services as we had done for John Graham on the Wells Fargo Building. A year later, in a joint venture with another Los Angeles firm, Parker and Zender, we designed the Bunker Hill Apartment Complex, including the thirty-story Bunker Hill Towers Building.

Scott: And ultrasonic testing was a very important part of those developments?

Nicoletti: Yes, ultrasonic testing was very important. Ultrasonic testing provided construction quality assurance for the field welding.

Did Not Identify Northridge-Type Problems

Nicoletti: Ultrasonic testing did not, however, identify conditions that led to the performance problems that surfaced after the Northridge earthquake in 1994. Tests by Professor Egor Popov at Berkeley with relatively light steel sections (16-inch-deep wide flange sections), and assurances by the steel industry, had given the structural engineering community confidence in the design of full-penetration welded beam-to-column joints in steel moment frames. The weld failures identified in southern California after the Northridge earthquake were a complete surprise to many structural engineers who had been using these details all over the world.

I served on an advisory committee of SAC Steel Project—the joint venture of the Structural Engineers Association of California, the Applied Technology Council, and California (now Consortium of) Universities in Research in Earthquake Engineering—that identified the issues and developed the scope for the subsequent investigations funded by FEMA. It was determined that the vulnerability of the welded joints was related to a number of factors, including: the depth of the beam, the thickness of the beam and column flanges, the type of electrode, and potential incomplete penetration of the initial weld pass at the backup bar.

SAC has issued several interim reports with recommendations for the repair and strengthening of existing joints and for the design of new joints. Los Angeles mandated the inspection and repair of welded joints within a given radius of the Northridge epicenter. However, building departments in other cities cannot force an owner to inspect his building in an area not affected by an earthquake or to strengthen undamaged joints. A few building owners in northern, as well as in southern, California have voluntarily had their buildings inspected, and in some cases have selectively strengthened the joints. The costs can be as much as $4,000 to $6,000 per beam-column joint, and most building owners would rather not be made aware of these problems in their buildings.

Scott: Is there any hope for the old joint details?

Nicoletti: The SAC investigations indicated that for many beam and column sizes, acceptable joints can be obtained if specific welding procedures are followed. Current building codes require prequalification of the beam-column joint details by testing to a specified inelastic joint rotation. The joints may also be prequalified by reference to prior test results performed with the same size members and with the same structural detailing and welding procedures. The steel industry and the testing laboratories can provide access to prior test results.

My current personal preference for the design of new steel moment frames is haunched beams designed so that the plastic hinge is developed in the beam at the beginning of the haunch and the weld to the column is always in the elastic range. Another alternative that seems to be attractive is the so-called "dog bone" configuration for the beams. This is a scheme where a portion of the top and bottom flanges of the beam are removed on each side of the web adjacent to the column, leaving the ends of the beam looking somewhat like a dog bone in plan. This weaker portion of the beam then acts as a "fuse" in that, under extreme lateral loads, it yields and protects the welded connection to the column.

100 California Street

Nicoletti: Our office has been involved in a number of major buildings in San Francisco. In addition to the Federal Office Building, which was field-bolted, we also designed this building that we occupied until a few years ago, 100 California Street. It was designed for Bethlehem Steel Company, and was field-riveted, although it had some shop welding. The building has offset columns—the exterior columns in one face are set off from the beam line. A torsion box was designed to take the torsion from the spandrel beam into the column. The spandrel beam is in the plane of the windows and the columns are offset. The torsion box had to be shop-welded, but the building was predominantly field-riveted.

Scott: Is that partly because it was done for Bethlehem Steel?

Nicoletti: The introduction of high-strength bolts for friction connections began to replace field riveting in the early 1960s. We had designed the building to be field-bolted, but Bethlehem had us redesign for field-riveting. I think that this was the last riveting crew that Bethlehem Steel had in the field, and this was probably their last job. It was the end of an era.[3]

The Embarcadero Center and Other Projects

Nicoletti: About 1965 we were approached by John Portman, an architect in Atlanta, Georgia. He also had an engineering department, and was going to do a major project in San Francisco, which became the Embarcadero Center. He wanted some consultation on earthquake design, so we started out with Embarcadero One.

Analysis and Design

Nicoletti: I was the project engineer and, working with Bill Chaw, we provided the criteria and did the computer analysis for them, and then John Portman did the design. Then we did Embarcadero Two and Three, and then the Hyatt Regency Hotel, pretty much on the same basis. We developed the earthquake criteria and did the analysis, and they would do the design and send it back to us. Then we would revise our model and analyze it again until we got compliance.

The seismic criteria that we developed for these projects would later be formalized as the IDR (inelastic demand ratio) methodology in the Tri-Service Manuals that we prepared for the U.S. Army Corps of Engineers, which I can discuss later. It consisted of monitoring component stresses and limiting the ratio of the calculated stress to the yield stress. The methodology was later improved and expanded as the linear elastic procedure in FEMA 273 and 310. We can return to those documents later.

On Embarcadero Four, with Andy Merovich, we actually did the total structural design for the frame, and did the analysis as well as the seismic design and the drawings. We did the same for the Bonaventure Hotel in Los Angeles, and also for the Portman Hotel here in San Francisco. The Embarcadero Four building, the last of the Embarcadero Center buildings, has some interesting features. It has a little wider column spacing than the other Embarcadero Center buildings. In our analysis we found that it was quite flexible in one direction.

3. Rivets were installed red-hot through pre-made holes in the pieces of metal to be connected. As they cooled, the rivets' length shrank and pulled the pieces of metal together. The high-strength bolt is strong enough that it can be so highly torqued with a wrench and tensioned that it accomplishes the same objective of tightly squeezing the connected pieces of metal together.

Eccentrically Braced Frames

Nicoletti: Even though it met the code, we found that the Embarcadero Four was too flexible in one direction. So we decided to stiffen it by introducing two eccentric braced frames within the moment frames. The eccentric braced frames were a new system that had been developed through research by Professor Egor Popov at the University of California. We had provided some consultation to Popov during this research period, so we were familiar with and quite enthusiastic about the potential capability to provide controlled stiffness. This is a chief feature of the eccentric braced frame—you can control the stiffness and provide good ductility.

Scott: Development of the eccentric braced frame design was an important advance, wasn't it?

Nicoletti: Yes. A concentric braced frame has very limited ductility. With a concentric braced frame, you only have one stiffness—for a given set of members you only have one elastic stiffness and limited ductility. Whereas with an eccentric braced frame, even for a given member size, you can have variable inelastic stiffness, depending on the amount of eccentricity. If we had put in concentric frames, they would have been so stiff that they would have taken most of the load, and we would have had to have very large bracing. With eccentric bracing, we could control the amount of bracing and the amount of additional stiffness, provide much better ductility, and make it compatible with the moment frame. This worked out very well.

Scott: So the eccentric braced frames design has inherent advantages over the regular concentrically braced frames?

Nicoletti: Yes. But one problem we had was that the code at that time did not recognize the eccentric braced frame, and San Francisco did not know what to do when we submitted the design. However, we were able to show that we could meet the code without the bracing, and then we would add the bracing for additional stiffness. You cannot make the building any worse by adding something, so they accepted it.

Scott: So even with the original amount of flexibility, which you thought was excessive, it would actually meet the code?

Nicoletti: There were no displacement requirements related to design seismic forces in the code at the time, although there are now. Since we met all of the force requirements, we were only making it stronger as well as stiffer.

Working with Blume on Seismic Codes

Scott: Is this a good place to discuss some of the other early work with John Blume, such as work on codes, including San Francisco's code, and the Alexander Building?

Nicoletti: Yes, they go together. I noted previously that I had learned very little about earthquakes until I started with John Blume, but I actually had very little contact with that area of his expertise until about 1948, when John was on a committee that was helping San Francisco write a seismic code. San Francisco's building code had no explicit seismic provisions until 1948. Before that, the state minimum requirements, the Riley Act and the Field Act, applied to buildings built in San Francisco. I helped John develop some examples to test the proposed provisions of the 1948 San Francisco building code.

Alexander Building, Separate 66, and the Blue Book

Nicoletti: The Alexander Building was one of the buildings that we tried the provisions on. John had done a lot of research on that building when he did his master's thesis at Stanford.[4] He had actually developed a mechanical model to represent the dynamic response of the Alexander Building. The model consisted of masses, springs, "dash pots" (viscous dampers), and so forth. The model could be shaken to simulate the response of the actual building. That model is now in the John Blume Earthquake Engineering Center at Stanford.

Later John was part of the northern California group, the Joint Committee of SEAONC (Structural Engineers Association of Northern California) and ASCE (American Society of Civil Engineers), San Francisco Section. The Joint Committee developed a building code—first it was a trial code. Then later, Blume was on the statewide committee of SEAOC that developed the first Blue Book in the late 1950s.

Scott: The Joint Committee's report and building code were referred to as Separate 66.[5] This was a major consensus-building effort by northern California structural engineers to resolve growing concerns about the cost and complexity of the emerging seismic requirements.

Nicoletti: Yes, Separate 66 was published by the SEAONC-ASCE Northern California Joint Committee. Later, about 1957, the statewide SEAOC Seismology Committee was set up, and it developed the first Blue Book,[6] which was published in 1959. The recommendations in the Blue Book largely became the seismic portion of the UBC (Uniform Building Code).

Scott: Would you say more about those code-writing efforts? What was involved in writing such a seismic code, and what were some of the key issues to be dealt with?

Nicoletti: Up to that time, almost all building codes treated earthquake design as a static phenomenon. You applied a certain static load—as you also did for wind—and then looked at a building's ability to withstand the forces. But many engineers, such as John

4. John Blume, "Period Determinations and Other Earthquake Studies of a Fifteen-Story Building," *Proceedings of the First World Conference on Earthquake Engineering*. Earthquake Engineering Research Institute, 1956. Blume and fellow student Harry Hesselmeyer produced their master's theses on the Alexander Building in 1934; Blume later received his Ph.D. from Stanford after revisiting and augmenting his earlier work.

5. Arthur W. Anderson, John A. Blume, Henry J. Degenkolb, Harold B. Hammill, Edward M. Knapik, Henry L. Marchand, Henry C. Powers, John E. Rinne, George A. Sedgwick, and Harold O. Sjoberg, "Lateral Forces of Earthquake and Wind," *Proceedings, American Society of Civil Engineers*. Vol. 77, Separate No. 66, April 1951.

6. Structural Engineers Association of California, Seismology Committee, *Recommended Lateral Force Requirements and Commentary*. Editions subsequently issued occasionally to support the updating of the UBC, beginning with the first partial edition in 1959 and first complete (with Commentary) edition in 1960.

Blume, John Rinne, and others, realized that earthquakes were really a dynamic phenomenon. A building is subjected to cyclic ground motion for a very short period of time, and the building's dynamic response to the ground motion sets up forces within the building.

So first, the Separate 66 group in San Francisco, and later the statewide SEAOC committee organized in 1957, tried to approach the issue from a dynamic standpoint. The Separate 66 group was convened primarily to propose a seismic code for San Francisco, although it was eventually published in the 1951 *ASCE Transactions*. The 1957 SEAOC Seismology Committee began to draft the first Blue Book and the provisions, which were later adopted by the UBC.

These efforts came up with the idea of applying the forces in an inverted triangle, to simulate the dynamic response of the fundamental mode, and looked at the reduction factor of the structure's overall design overturning moment, the so-called "J" factor. They introduced a series of formulas that would allow you to design a building pretty much with static forces that would nevertheless try to simulate the dynamic response of the building.

Scott: So engineers could use traditional analytical procedures for static loads to represent the dynamic response of buildings?

Nicoletti: Yes. This was an early pseudo-dynamic approach to earthquake design. It used static design, but tried to build in some considerations that would help take care of the dynamics. Since then, of course, building codes have progressed along that way pretty much until today. Building codes are pretty much the same, except that with the advent of computers, dynamic analysis became feasible and now is becoming more and more common. The building code actually requires dynamic analysis for certain irregular buildings and certain unusual buildings.

Pioneering Dynamic Analysis: Work in New Zealand

Nicoletti: The Blume firm is considered a pioneer in the use of dynamic analysis for buildings. I think our first venture into this field was about 1955 or 1956. We were selected by the City of Auckland in New Zealand, to help them develop a new building code. Actually it was a building code essentially for the entire country of New Zealand. Up until that time, they had a six-story limit on all buildings. We helped them revise the code so they could build larger buildings.

Then Auckland said, "We want you to help us design our city administration building." They had a nineteen-story building in mind, and we helped them with it. I was the project engineer on that building. I did not do any of the design, but I helped them establish criteria, and I reviewed their design, and worked with them. Then about 1956, when the design was pretty far along, the Dominion Laboratory in Wellington—part of the government laboratory in New Zealand—had acquired an analog computer. One of the research engineers there contacted us and said he would like to have us work with him in making a mathematical model of the building and subjecting it to ground motion in the computer. As far as I know, this was the first time this was done for a building.

Scott: Was that to be done as kind of a scientific and intellectual experiment, or was it also going to have some more directly practical application? Did they think the results might motivate them to reconsider some aspects of the design?

Nicoletti: It was intended pretty much as an experiment, but also it was recognized that the experiment would either confirm or deny what we had recommended for the design of the building. So we acquired all the available records of major earthquakes in the U.S. of magnitude 6.5 or greater. There were not very many of them at that time, maybe six or so records at that point, and we "normalized" them. In other words, we got them to the same maximum amplitude, then we sent them down to New Zealand. We took the nineteen stories and lumped the mass of some of the stories, because the capacity of the computer was very limited. We had about six lumped masses, and we essentially performed a two-dimensional dynamic analysis of this lumped mass model with an analog computer, which instead of numbers gives you a graphic trace. That was our first dynamic analysis of a building, and as far as I know it was the first one of an actual building that had some practical application. Eric Elsesser, who worked in our office prior to starting his own firm with Nick Forell, helped Blume in preparing the ground motion and the building model.

Scott: How did it come out?

Nicoletti: It came out fine, we were happy with the results.

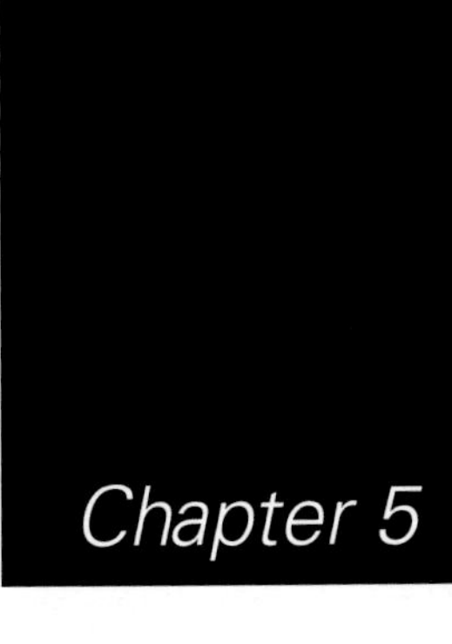

Post-San Fernando Work

The Navy had a large inventory of buildings, so they were looking for a rapid seismic evaluation technique.

Scott: Your discussion of the Embarcadero Center and eccentric braced frames has taken us chronologically to about 1970.

Nicoletti: Yes, and soon afterward we had the San Fernando earthquake on February 9, 1971, which was a big stimulus to earthquake engineering. The collapse of the Veterans Administration Hospital particularly spurred a lot of activity.

Scott: Yes, the resulting VA program was an important development.

Nicoletti: The Veterans Administration embarked on a program of seismically strengthening their hospitals nationwide, and of course other government agencies also began to look critically at their hospitals. Our firm was one of several selected by the Veterans Administration. We evaluated and strengthened several Veterans Administration hospitals throughout the country.

Shortly after the San Fernando earthquake, we were approached by the Navy. They decided they wanted us to evaluate two of their hospitals in hazardous areas. One was in Oakland, California and the other was in Charleston, South Carolina. I went to Washington to negotiate that contract, and found that the Navy was anticipating eventually going to have to look at their other facilities, such as shipyards and other naval facilities.

Rapid Evaluation Technique

Nicoletti: The Navy had a large inventory of buildings, so they were looking for a rapid seismic evaluation technique. I told them I thought we could come up with one. After we completed our contract on the Veterans Administration hospitals, we were asked to provide a proposal to develop a rapid evaluation technique for the seismic vulnerability of such buildings as might be found in naval facilities. We submitted a proposal, which was accepted. We started on the first phase of developing a rapid evaluation technique, and I had an idea of where I wanted to go, but was not quite sure of how we were going to get there. We started doing some preliminary work.

Atomic Energy Commission

Nicoletti: As we discussed earlier, our company had been involved for over ten years with what was then the Atomic Energy Commission at their test site in Nevada, near Las Vegas. In the process, we had developed an expertise and a lot of experience in the response of buildings. We decided to try to use this experience in developing the rapid seismic evaluation technique for the Navy. Sig Freeman, who was working for us then, had been involved in the Las Vegas work, and he was instrumental in developing the technique we finally adopted.

It is a semi-analytical and semi-graphical technique that is quite rapid in establishing the vulnerability of rather simple one-story and two-story buildings such as are found on naval facilities. This technique was accepted by the Navy, and became their standard method, and is still in use. Other people, of course, have also used it.

The main element is a response spectrum, which is sort of the signature of an earthquake. It is a graphical representation of the response of buildings to the earthquake. It relates the period of a structure to what is called the spectral acceleration, which is a function of the force that the building would see in an earthquake. Our technique treated the response spectrum as a demand. Superimposed on this, we would plot the capacity of the building as a function of spectral acceleration and period, starting with the elastic capacity; and then as members yielded, the change in period and the inelastic capacity of the structure after yielding. If our capacity curves exceeded the demand curves, then the building was okay. If the capacity curve fell below the demand curve, then the building was in trouble. Essentially, the technique combined some simple analysis of the structure with a graphical representation of the response spectrum. It is an approximate technique, but we confirmed its validity with more rigorous analysis, and found that it was quite acceptable.

The underlying basis of this was some of the work that John Blume had done, dating back to the early 1940s, on the "reserve energy" technique.[7] In other words, it was based on the conservation of energy. That's the technical basis for the procedure. Following our development of the technique, the Navy started issuing contracts for evaluating their facilities. Over a period of about ten years, we evaluated 300 or 400 structures at Bremerton, Washington, at Mare Island, California, and Charleston, South Carolina. Those were the ones we did for the Navy. Other firms did other locations.

Using the Rapid Evaluation Technique

Nicoletti: The procedure was first to get the drawings of the building. From those we could identify the primary lateral load resisting system. Then we would visit the building and confirm that it was built in accordance with the drawings. We verified that they hadn't left out bracing, or made some modification that would change the structural system. Then we would come back to the office and do our evaluation.

One of the products of the evaluation was an estimate of the damage for any given earthquake, which was obtained graphically from the plot. Every earthquake has a unique acceleration response spectrum, which is somewhat similar to a power density spectrum, that indicates the energy content at various frequencies. The earthquake response spectrum is an index of the maximum response of structures of various natural periods or frequencies. Standardized response spectra have been developed to represent the probable response demand on structures for various levels of ground motion and for different site conditions.

In our rapid evaluation procedure, these standardized spectra were used to represent the various demand curves. The capacity of the building was determined by simple static pushover analysis and plotted in terms of base shear versus roof displacement. These coordinates could then be converted to spectral acceleration versus fundamental period, and superimposed on the response spectrum. From this graphical consideration of capacity and demand, the potential damage for that particular ground motion could be estimated.

Methodology Confirmed by Loma Prieta Experience

Nicoletti: There is an interesting follow-up to this. I mentioned that we did something like a hundred buildings at Mare Island. All told, we did 300 or 400 buildings for the Navy at various locations. Some of the Mare Island buildings were damaged in the 1989 Loma Prieta, California earthquake, so the Navy asked me to go up and look at those buildings. I found that the damage was in very close compliance with the evaluation we had done. In other words, we had been able to predict which buildings were most vulnerable, and those were the ones damaged. This pretty well confirmed our evaluation method. Unfortunately, before the earthquake, the Navy had not done anything about the evaluations, and they have subsequently closed Mare Island as a naval base and turned the facilities over to the City of Vallejo.

Scott: So your long association with the development of methods to seismically evaluate

7. Blume, John A., "A Reserve Energy Technique for the Earthquake Design and Rating of Structures in the Inelastic Range," *Proceedings of the Second World Conference on Earthquake Engineering*. Vol. II. Tokyo, Japan, 1960.

existing structures started out back about 1971 at the time of the San Fernando earthquake?

Seismic Design Manuals for the Army

Nicoletti: That's right. We started out with the Veterans Administration, and then the Navy hospitals and the rapid evaluation. And now we should discuss the Tri-Service Manual. The Tri-Service Manual and the two supplements to it are Department of Defense manuals for all three services. They were prepared under contract with the Army Corps of Engineers.

The Tri-Service Manual[8] was sort of an interpretation of the building code, with design examples and commentary. In other words, it was something of a cookbook approach to seismic design. That was what they wanted, because they had engineers in other parts of the country who were not familiar with seismic design, and yet there were requirements for seismic design.

8. Called "Tri-Service" because it was published by the Army, Air Force, and Navy (with the Marine Corps facility design function part of the Navy publication), *Seismic Design for Buildings* was originally produced in 1966 and has been revised and re-published in subsequent editions. The same document has three document numbers for the three military services: Army TM 5-809-10; Navy NAVFAC P-355; Air Force AFM 88-3. Part textbook and part guideline or surrogate building code, the book provides the criteria for projects for these services, but has also been widely referred to by engineers for other applications.

Scott: When you say "an interpretation of the building code," do you mean the Uniform Building Code?

Nicoletti: Yes. We helped the Army update this manual in 1976, and then revised it again for them in 1982. The Tri-Service Manual has been superseded by TI 809-04, which I will discuss with the more recent manuals prepared for the Department of Defense.

Supplements for Essential Facilities

Nicoletti: In 1985, the U.S. Army Corps of Engineers asked us if we could develop supplements to the Tri-Service Manual for essential facilities; that is, facilities that need to be designed in excess of the UBC requirements. These are facilities such as communications facilities, hospitals, fire protection, and so forth. They were interested in maintaining essential facilities and essential functions after severe earthquakes.

Sig Freeman and I worked together and developed a seismic design manual for essential facilities.[9] It starts with the basic requirements of the code, and then provides special seismic requirements. Here again, we planned some new techniques, providing two different methods for evaluating essential facilities. These techniques both involved the use of an elastic spectral response analysis to represent the inelastic response of buildings. These two methods are called the IDR (inelastic demand ratio), and the capacity spectrum method.

9. *Seismic Design Guidelines for Essential Buildings.* Army TM-809-10-1, Navy NAVFAC P-355.1, Air Force AFM 88-3. Chapter 13 Section A, 1986.

In the IDR method, for each structural component, the IDR is determined from analysis as the ratio of the calculated elastic displacement to the yield displacement for that component. Since the analysis is elastic, the ratios of the moment or shear responses—which are proportional to the displacement ratio—are generally used as they are more readily available in the analysis. The structural response is acceptable if the calculated IDRs are less than the values tabulated in the document. The technical basis for this lies in the fact that the combined elastic and inelastic displacement response of most buildings to an earthquake is no greater or even less than if the building had remained totally elastic. (For buildings with shorter fundamental periods—generally less than about one-half second—this relationship does not apply without some adjustment). This procedure was an extension and a refinement of the procedure I had developed for the seismic analysis of the Embarcadero Center buildings we had done for John Portman.The capacity spectrum method is a refinement of the rapid evaluation technique we developed for the Navy. After the initial elastic analysis, the member responses are normalized, or prorated, to the values that result in initiation of yielding in the most highly stressed member. Plastic yield hinges for that member are introduced in the structural model, and the analysis repeated. This process is reiterated until a collapse mechanism occurs in the model. The responses are converted to spectral values and are plotted by superposition of the iterative analyses to construct a capacity curve that is graphically compared with the demand as represented by the response spectrum.

The capacity spectrum method is considered to be a more realistic approximation of the inelastic response, but the IDR method is quicker and more easily implemented, and thus has gained greater popularity for new construction, as well as for the evaluation of existing structures. This supplement to the Tri-Service Manual is entitled "Seismic Design Guidelines for Essential Facilities," and was published in 1986.

Upgrading Existing Buildings: Breaking New Ground

Nicoletti: In 1988, Sig Freeman and I prepared a companion document entitled "Seismic Design Guidelines for Upgrading Existing Buildings."

These two supplements were breaking new ground, especially the second one for existing buildings. This was one of the first manuals to treat existing buildings with other than just the basic code approach used for new buildings. Both manuals have been used extensively by other engineers, even though it is recognized that their methods are simplified rather than rigorously accurate. I think they were a great improvement over the procedures in the prior codes.

The current code approach to seismic design relies on inelastic response to allow structures to be designed for forces much less than those associated with the ground motion. The code recognizes this by using a response reduction factor. This takes the realistic earthquake loads and reduces them down to design loads. The response reduction factors in the Uniform Building Code vary from about 4 to 12. What these factors do is arbitrarily reduce the realistic earthquake loads, which are then used on a global basis—the whole structure is designed with these reduced loads. By contrast, our methodology retains realistic earthquake loads. It then looks at the response of the building and evaluates the response of individual members. Thus, we can be selective in the response modification. In reality, we are recognizing that some structural components can sustain larger inelastic deformations than others.

Scott: You would reduce it less for some members than for others?

Nicoletti: Right. We would recognize that columns in framed structures are very vulnerable. We would reduce the response of columns used in design calculations by a small amount, and maybe reduce the corresponding response of the beams by a larger amount.

Scott: That would really depend on their roles in supporting and holding the system together under earthquake forces?

Nicoletti: Yes. So in the seismic rehabilitation of a building, our elastic analysis would first of all identify the critical members, because those would be the ones that would show up as hot spots. Then we would look at the response with the IDR criteria and, if necessary, modify the design of the structure. Again, as I said, this is not a rigorously accurate method, but I think it is a big improvement over what the prior codes provided. This approach was adopted by FEMA in some of their recent documents, FEMA 273 and 310, which we should also discuss.

Scott: Yes. I know that you were involved with a number of the FEMA documents through FEMA and ATC.

Nicoletti: The National Earthquake Hazards Reduction Program (NEHRP) was enacted as public law in 1977, a few years after the 1971 San Fernando earthquake. NEHRP called on all public agencies to assess the seismic risk in all government-owned or government-leased buildings. Only a few government agencies actually took action in response to this legislation, as there was no required time frame specified for the action. In November 1990, following the 1989 Loma Prieta earthquake, Congress reauthorized NEHRP with Public Law 101-614, and in December 1994, President Clinton signed Executive Order 12941, which required all federal agencies that owned, leased, or financed buildings to assess the seismic vulnerability of their inventory and report their findings to FEMA by December 1998.

Screening and Evaluation Procedures

Nicoletti: Shortly after the Executive Order was issued, the U.S. Army Corps of Engineers contracted with URS to develop manuals that they could use to comply with the Executive Order. I was the principal author of the two manuals—EM 1110-2-6052, February 1996, *Screening and Evaluation Procedures for Existing Civil Works Buildings;* and EI 01S103, March 1997, *Screening and Evaluation Procedures for Existing Military Buildings.* These were based in a general way on FEMA 154 and 178,[10] but

10. FEMA 154, *Rapid Screening of Buildings for Potential Seismic Hazard*, 1988; and FEMA 178, *Handbook for the Seismic Evaluation of Existing Buildings*, 1992.

were adapted to apply specifically to military buildings.

The original purpose of the manuals was to provide guidance to perform the necessary evaluations for Army engineers, who may have little or no experience in seismic engineering. As it turned out, engineers out of our office performed all the screening and evaluations at selected Army installations throughout the U.S. as well as at a number of Navy and Air Force installations.

I was asked to review the evaluations and provide guidance, particularly since many of the older buildings were not designed for seismic forces and contained structural systems that were strange to our younger engineers.

Updating the Tri-Service Manual

Scott: You mentioned earlier that you also prepared a document to replace the Tri-Service Manual.

Nicoletti: Yes. In 1996, the Army Corps of Engineers authorized URS to prepare two new manuals, one to replace the Tri-Service Manual for the seismic design of new military buildings, and the other for the seismic evaluation and rehabilitation of existing buildings (TI 809-04, 1998, *Seismic Design for Buildings*, and TI 809-05, 1999, *Seismic Evaluation and Rehabilitation for Buildings*). Again I was the principal author for both of these manuals.

Unlike the prior screening and evaluation manuals that were prepared specifically for use on Army buildings, these new manuals were for the use of all three services and were reviewed and critiqued periodically during preparation by representatives of all three services. The design manual referenced FEMA 302 provisions[11] for ordinary buildings with a life safety performance objective, and modified the provisions of FEMA 273[12] for the design of buildings with enhanced performance objectives. The Tri-Service existing buildings manual referenced FEMA 310[13] for seismic evaluation, and modified the procedures of FEMA 273 to apply to rehabilitation.

Scott: It appears that the Department of Defense is taking NEHRP very seriously.

Nicoletti: Yes. I believe that they have been more active in following NEHRP than the other federal agencies and, of course, they have a lot of buildings. Each one of the military services has about 30,000 buildings.

As I mentioned earlier, we had helped the Naval Facilities Engineering Command, NAVFAC, perform seismic evaluation of their facilities in the continental U.S. in response to President Clinton's Executive Order 12941. Apparently Navy personnel stationed overseas asked why their facilities had been omitted. The Naval European Command decided that probably the most vulnerable facilities were the housing units, built to local codes and leased by the Navy for their personnel. Funding was appropriated and the Atlantic Division of NAVFAC was directed to oversee the evaluations. In early 2001, I was contacted by the

11. FEMA 302, *NEHRP Recommended Provisions for Seismic Regulations for New Buildings and Other Structures*, 1998.
12. FEMA 273, *NEHRP Guidelines for the Seismic Rehabilitation of Buildings*, 1997.
13. FEMA 310, *Handbook for the Seismic Evaluation of Buildings: A Prestandard*, 1998.

Atlantic Division for suggestions as to how the evaluations should be performed. I informed them that, in my opinion, the screening and evaluation procedures developed by FEMA may not be applicable because of the differences in construction. They agreed and we decided that it would be advisable to do a quick reconnaissance to determine the typical structural systems and to develop applicable evaluation procedures.

Evaluating Housing in Naples for the Navy

Nicoletti: The initial location was the Naval Support Facility in Naples, Italy and the leased housing was located in the small towns bordering the city. I conducted the reconnaissance with an engineer from the Atlantic Division and it turned out to be quite interesting. The housing units are typically two- or three-story buildings with a basement. There are generally two apartments per story, with parking in the basement. The units are all less than ten years old and have remarkably similar structural systems. The basic system is a reinforced concrete frame. At the exterior walls, the frame is infilled with hollow clay tile. The concrete columns are constructed to the soffit of the floor or roof beams and a temporary plywood platform is erected to support interlocking hollow clay tile, placed with gaps to form the floor or roof beams. Beam and slab reinforcement is installed and concrete is placed in the gaps and over the tile units. This provides a flush ceiling requiring only a finish coat of plaster.

We later discovered that this structural system, with minor variations, is common throughout post-war Europe. It makes optimum use of common local materials (i.e., clay, cement, and aggregates) and provides a sturdy building with good insulation and weather resistance.

The units in the Naples area were designed to the Italian building code that prescribed elastic design to about .13g seismic forces. Our evaluation was to be based on about .40g. Complete drawing sets were usually available either from the owner or the local building department. Additionally, copies of the building permit were also available. The building permits were quite helpful as they contained statements from a soils engineer regarding the foundations and from a structural engineer regarding the basis of design. From these documents, we could determine the shear and moment capacity of the columns. I proposed a preliminary evaluation procedure whereby the buildings would be rated as Category A (acceptable), B (needs further study), or C (needs retrofit). The evaluation focused on the reinforced concrete columns; the hollow clay tile was assumed to add mass but no resistance. The columns were evaluated on the basis of *m* values derived from FEMA 273. The results of the evaluation appeared to be rational and the procedure was extended for the evaluation of similar buildings at U.S. Naval stations in Sicily, Greece, Crete, and Spain.

California Hospital Act

Scott: California's landmark Hospital Seismic Safety Act was passed in 1972 as a result of hospital damage in the San Fernando earthquake. The act's chief goal was to ensure that new hospitals built in California would be able to function effectively after an earthquake in the immediate area.

Nicoletti: Let me tell you a story about the Hospital Act. Right after the San Fernando earthquake, we got a call from the Office of the State Architect in Sacramento; now it's called the Division of the State Architect. They wanted to know if we could help them write a code for California hospitals. So I met in Los Angeles with some representatives of the Division of the State Architect from its Sacramento, San Francisco, and Los Angeles offices—Jack Meehan, Bob Benson, and Don Jephcott, respectively. Together we came up with many of the provisions that are currently in the state hospital code.

Those Hospital Seismic Safety Act provisions were prepared with the understanding that they were to be only interim provisions, and that within a month or so something better would be done. Well, that was over twenty years ago, and we are still using those provisions. They have been modified a little bit, but at the time they had in mind doing something much more elaborate and much more sophisticated. They have not gotten around to it yet. As a matter of fact, in recent years they have pretty much adopted the Uniform Building Code, with an Importance Factor of 1.25.

Scott: I had not realized that the code for hospitals is now basically so similar to the Uniform Building Code.

Nicoletti: It is now, yes, in terms of basic design approach, although plan review and construction quality assurance procedures are much more strict under the Hospital Act. What we did twenty years ago was to increase the design factor, and as an alternative to this increased factor we would permit a dynamic analysis to a site-specific earthquake determination. That still remains in the code.

Scott: So the Uniform Building Code has kind of caught up with the hospital regulations?

Nicoletti: Right.

Code Development

Out of the fifty or sixty people present at this meeting, there was not a single voice that gave any support to the proposed ATC-3 code—they were all against it because it upset the status quo.

Nicoletti: Starting right after the San Fernando earthquake I began to become involved with code development. I was on the Seismology Committee of SEAONC—the Structural Engineers Association of Northern California—and later I was on the statewide SEAOC Seismology Committee. About that time, SEAOC formed the Applied Technology Council (ATC).

Scott: What was the main purpose of setting up ATC?

Nicoletti: ATC was created because it was recognized that working on code provisions only on a volunteer basis was not very efficient or effective. So ATC is a nonprofit vehicle to obtain funding from public agencies so that more focused code development work can be done. ATC initially paid the participants at least a nominal sum, I think $30 or $40 an hour. Not exactly the going rate for structural engineers, but still paying something. Currently ATC pays $75 to $100 an hour, depending on the contract and the nature of the services required.

Scott: ATC also provides for some paid staff, whereas previously under the all-volunteer system I guess there had hardly been any paid staff work at all.

Nicoletti: That's right, it had all been done on a volunteer basis.

ATC-3-06: New Seismic Code and Consensus Process

Nicoletti: One of the first major tasks of ATC was to prepare ATC-3-06,[14] which was the development of a new seismic code, funded by the National Science Foundation and the National Bureau of Standards. The National Bureau of Standards was later re-named the National Institute of Standards and Technology, but is still headquartered in Gaithersburg, Maryland. ATC-3 was quite a radical departure from the UBC methodology of that time, although the force levels were selected to be compatible with the 1973 UBC. ATC-3 was a five-year effort by a large group of people, including John Blume. I was not directly involved in ATC-3, except that I was on the ATC board of directors while ATC-3 was being done, and I was president of ATC when it was finally finished. The 1988 UBC was the result of another five-year effort by SEAOC to adopt many of the provisions of ATC-3.

14. *Tentative Provisions for the Development of Seismic Regulations for for Buildings.* National Bureau of Standards and the National Science Foundation, 1978. The "-06" nomenclature had to do with the number of the draft that became the final version—there was no connection to the '06 of 1906 San Francisco earthquake fame.

Scott: So you were pretty well aware of ATC-3, even though you were not directly involved in working on it?

Nicoletti: Yes, I was directly involved in what they called the consensus process, which came after the document was finished by ATC. That was a very painful operation. We had to deal with other code bodies, and the materials people were really a pain in the neck. They were all jockeying for position, and I hadn't even realized they existed. I was president of ATC at the time, so I was personally involved in the process.

Scott: Would you go into that process a little? The processes of code development are interesting and important, but the subject does not get treated in the technical engineering literature very much.

Nicoletti: Traditionally, SEAOC, through its Seismology Committee, prepared provisions that were given to ICBO (International Conference of Building Officials). ICBO was the body that published the Uniform Building Code. SEAOC would prepare the seismic recommendations, its *Recommended Lateral Force Requirements and Commentary*, the Blue Book, in advance of each edition of the Uniform Building Code, which comes out every three years. ICBO would take these recommendations, and using its committees, would go through the consensus process and eventually publish the provisions in the UBC, which could then be adopted by cities and counties to become local ordinances. So SEAOC had been pretty much sheltered from the consensus process by ICBO.

My first exposure to the consensus process was on this ATC-3 project, when we had actually made a drastic change in the building code. We submitted it to the National Bureau of Standards. I think they were handling the consensus process. Later on, this was done by the National Institute of Building Sciences, NIBS, through the organization set up within it called BSSC, the Building Seismic Safety Council. The National Bureau of Standards arranged a meeting that was held in Silver Springs, Maryland in 1978, attended by Rol Sharpe, Ron Mayes, and me. There were representatives of all of the building code organizations, and of course all the materials people—brick or masonry, wood, concrete, and steel. Out of the fifty or sixty people present at this meeting, there was not a single voice that gave any support to the proposed ATC-3 code—they were all against it because it upset the status quo.

Scott: They were all against what was recommended by the ATC project?

Nicoletti: Right. There was not a single statement in favor of those provisions, except for the three of us representing the ATC project. Everybody else was against it. They were all suspicious that they were going to lose position.

Scott: You mean, for example, that people representing one material were afraid their manufacturing and construction industry might lose its competitive position as compared with another industry based on another material?

Nicoletti: Right. So this led to the formation of BSSC, funded by FEMA.

BSSC and NEHRP: Toward a National Consensus Code

Scott: Briefly describe the Building Seismic Safety Council (BSSC)—I believe it was part of a long-term push toward a national consensus on seismic design standards.

Nicoletti: Yes. BSSC is a council that represents all the interested organizations in the construction field: engineers, architects, building officials, construction unions, and representatives of the materials organizations. The BSSC board of directors consists of representatives from many of these organizations. I have served on the board, representing EERI. Involving all of these organizations throughout the development of seismic provisions has worked well for BSSC and for FEMA, the federal agency with the responsibility to develop the provisions.

For ATC-3, BSSC embarked on what became almost a ten-year consensus process. They selected engineers in various parts of the country to design trial buildings, to be estimated and compared with the current code, and so forth. Very little change was made to the ATC provisions. They finally reached consensus, with a few minor changes here and there.

Scott: They ended up not making many changes in the ATC code provisions that previously everybody else had been opposed to?

Nicoletti: Yes, they had over-reacted to the new methodology, without realizing that it was designed to have about the same impact on the buildings that were built, but make the seismic design process more rational. So the ATC-3 provisions became what is now known as the NEHRP provisions.[15] The NEHRP provisions are revised every three years, and I have been involved in several of the modifications.

Scott: And this process has resulted in a gradual movement toward a consensus on code standards?

Merger and the First National Consensus Building Code

Nicoletti: Yes, the NEHRP provisions were the beginning of a national consensus code. Prior to being adopted by the International Building Code (IBC), they were adopted by the Southern Building Code, and also by the City of New York.

Scott: Say a word or two about the developments that led to the IBC.

Nicoletti: Various building official organizations previously published three different model codes—the Uniform Building Code, the Southern Building Code, and the Building Officials and Code Administrators (BOCA) National Building Code. Each of those codes was written as if it were a national code, but they were generally adopted in different regions, with the UBC being the code in California and the Western U.S., the Southern Building Code in the South, and the BOCA code in the Midwest.[16] They merged to form the International Code Council, ICC, to publish the International Building Code, the IBC. The IBC 2000 is the first national consensus building code in the U.S., and its seismic provisions were based on the 1997 NEHRP provisions.

FEMA is funding BSSC to continue to publish updated versions of the NEHRP provisions. I also understand that SEAOC will continue to publish their Blue Book, but the ICC has not committed itself to either SEAOC or BSSC for future updates of the IBC, although the people who update the provisions will probably be the same ones who have done so in the past, regardless of their affiliation.

Scott: Can you summarize the main ways in which the Uniform Building Code differed from the new IBC?

Nicoletti: Although the 1988 UBC was based somewhat on the NEHRP provisions, it retained what is known as the working stress method, rather than the ultimate strength method. That is just a difference in philosophy—or rather not so much in philosophy as in the design allowable stresses themselves. It was a very simple numerical difference, but it affected many of the equations and detailing provisions in the UBC.

The 1997 UBC adopted the ultimate strength approach and finally was in close compliance with the NEHRP provisions. This should not have been too surprising, as many California engineers participated in the preparation of both documents.

---

15. FEMA 368, *NEHRP Recommended Provisions for Seismic Regulations for New Buildings and Other Structures*, FEMA 368 (Part 1, Provisions) and FEMA 369 (Part 2, Commentary), 2000 edition (March 2001).

16. The BOCA, or national code, was distinct from the original National Building Code, begun in 1905 by the insurance industry, which was the first model building code in the U.S. intended for national application, and which ceased publication in the 1970s.

Scott: What is the principal difference between the working stress and the ultimate strength approaches?

Nicoletti: Traditionally, structural materials have been specified on the basis of allowable stresses. And the allowable stress for most materials is generally about 60 percent of the yield stress. That is sort of a working factor of safety. For earthquake design, ductile yielding is used to reduce the seismic forces, and it becomes much easier to work with the yield stress rather than the allowable stress. You can compensate for it by having other load factors, or other coefficients that allow you to treat normal operating loads with a lower stress. That was the basis of an allowable stress—you wanted operating loads to be at a safe stress that could be achieved without any problem.

So the code started out with everything being on allowable stress, and when ATC-3 was developed, they recognized the advantage of going to the yield stress (the ultimate strength) approach for seismic provisions, and ATC-3 was written on that basis. The 1988 UBC, as I said, adopted pretty much the ATC-3 philosophy, but used the allowable stress approach rather then the ultimate strength approach. The 1997 UBC finally adopted strength design and is very similar to the NEHRP provisions. As I indicated earlier, the ATC-3 provisions were eventually adopted by BSSC. They were then tailored to become the first edition of the NEHRP provisions as FEMA 95 in 1985. I have also indicated earlier that FEMA intends to update these provisions every three years. The last edition of the UBC was in 1997.

Seismic Codes Originally Developed by SEAOC

Scott: The purpose of creating BSSC was, at least in part, to try to set up a structure for the consensus-building process, when it had become clear that a lot of consensus-building needed to be done. Was it largely a matter of it being hard for the non-Californians to accept a California-based code? Was the consensus building needed in part to overcome this resistance?

Nicoletti: California obviously has the biggest need for a seismic code, and most of the provisions traditionally have come from SEAOC, most of which are found not only in the UBC, but also in other codes around the country, and even in codes in foreign countries where they adapted their provisions from the seismic provisions in UBC.

ATC-3 was essentially written by California earthquake engineers, with a few exceptions, such as people like Nathan Newmark from Illinois, Glen Berg from Michigan, and a few other people who were pretty much pioneers in earthquake engineering, even though they were not themselves located in earthquake country.

Basically, however, ATC-3 was a code developed in recognition of California's requirements, but it was also developed to address the seismic hazard in other parts of the U.S. I believe that many of the initial objections to ATC-3 were of the "if it ain't broke, don't fix it" variety. Many saw no need for new seismic provisions and were concerned about what it would do to the status quo of their business or professional practice. I believe that FEMA, BSSC, and NSF have done a remarkable job in raising the level of public consciousness regarding the seismic hazard in the central and eastern U.S.

In 1988, ATC developed guidelines and manuals for the U.S. Post Office for the evaluation and seismic retrofit of their buildings. The Postal Service has 37,000 buildings, many of which are very small, but they wanted to do an evaluation of them. ATC selected evaluation procedures similar to ATC-22[17] and also provided guidelines for the retrofit. In developing the manuals for the Post Office to evaluate their buildings, ATC has generally adopted the NEHRP provisions.

I chaired the project engineering panel as the project went through four phases and was completed in 1991 as ATC-26. Manuals were prepared for the seismic evaluation and strengthening of the buildings, and also for training Post Office personnel to do a preliminary walk-through after an earthquake. The program also included demonstration projects to test the guidelines. Representative postal buildings were selected from throughout the U.S. for evaluation of the ATC guidelines.[18]

ATC-34—A Joint NCEER-ATC Project

Nicoletti: *Joint Study of R factors and Other Critical Code Issues* was a joint project of ATC and the National Center for Earthquake Engineering Research (NCEER).[19] I was on NCEER's Scientific Advisory Board from 1993 to 1999. NCEER provided the funding and ATC provided the project management.

I was on the project engineering panel, chaired by Ian Buckle, and it included engineers and academics from diverse geographic locations throughout the U.S. The original assignment of ATC-34 was to look at the R factors. The R factors reduce the design loads from what we actually expect in the design earthquake to load values that calibrate with the way the capacities are calculated. As the project developed, it was obvious that we could not divorce the R factors from all the considerations that go into designing buildings inelastically, because we know buildings won't stay elastic in a big earthquake. The attempt was to adequately predict inelastic behavior. These considerations included response reduction factors, ductility, the methods of analysis, and so forth, which are all related. So the scope of the project was broadened to look at all of these things.

The revised scope of work was closely tied in with concurrent work by SEAOC under their Vision 2000 Project. The ATC project provided some of the technical background and research that SEAOC could use as a basis for future code recommendations.

The document included a critical look at the various seismic provisions currently in use and identified their strengths and weaknesses. It

17. ATC-22, *Handbook for Seismic Evaluation of Existing Buildings* (preliminary). Applied Technology Council, Redwood City, California, 1989.

18. A series of reports: ATC-26-1, ATC-26-2, published by the Applied Technology Council, Redwood City, California.

19. ATC-34, published by FEMA. 1994; NCEER has since been reorganized and re-funded by NSF as the Multidisciplinary Center for Earthquake Engineering Research, MCEER, and is still headquartered at the University at Buffalo, State University of New York.

also provided a new format for the next generation of seismic provisions, with a revised set of R factors for very simple buildings. These R factors could be used for preliminary design of other buildings, but the final designs would have to be confirmed by some sort of component-based linear or nonlinear analysis. Two additional phases of the work were proposed, to develop new R factors and approximate nonlinear analyses, these to be confirmed and calibrated against more rigorous analytical procedures. No additional funding was available, however, to support such work.

Vision 2000

Nicoletti: The Vision 2000 Project[20] was funded by the California Office of Emergency Services. I served on the technical committee, chaired by Professor Vitelmo Bertero.

The Vision 2000 Project was intended to provide "vision," or guidance, for those who will be drafting the Blue Book recommendations in the year 2000 and beyond. There are acknowledged weaknesses in the current Blue Book, particularly the R factors. While the Vision 2000 Project will thus give guidance to help in revising the Blue Book, the updating process itself will probably also take other and different approaches.

The Vision 2000 Project was completed in 1995, and provided direction that envisioned the escalation of the seismic provisions for several progressive performance levels, with minimum requirements for the associated analytical procedures. This would vary from prescriptive (with no design analysis required) provisions for simple buildings with life safety as the only performance objective, to nonlinear inelastic analyses required for complex essential facilities that should remain functional during and following a major earthquake.

20. *Vision 2000, Performance-Based Seismic Engineering of Buildings.* Structural Engineers Association of California, 1995.

Scott: In due course, Vision 2000 could be reflected as recommendations in the Blue Book, be incorporated in the building code, and show up in the standards of practice?

Nicoletti: Yes. It should be remembered that this document was prepared concurrently with FEMA 273, the engineering guidelines for the rehabilitation of existing buildings, and the updates to FEMA 302, which is the 1997 edition of the NEHRP provisions for new buildings, and with some of the same personnel working on all three projects. Many of the innovations proposed in Vision 2000 were incorporated in FEMA 273, and to a much lesser extent in FEMA 302, because that was an updated existing document. Since the seismic provisions in the IBC 2000 are based on FEMA 302, however, the IBC will not reflect these innovations until some future issue.

Scott: How will the outcome of the ATC and SEAOC efforts relate to the Blue Book? Will this produce a new-generation and more sophisticated version of the Blue Book?

Nicoletti: The Vision 2000 document is intended as recommendations to the SEAOC Seismology Committee, which publishes the Blue Book. Many of the ideas in the document have already been incorporated in subsequent editions of the Blue Book.

The R Factor

Scott: Would you discuss the R factor and indicate why it is a point of particular concern?

Nicoletti: We have been designing buildings at force levels that are far below the elastic force levels associated with recorded earthquake ground motion. Most of our earthquake provisions have come from experience. We know from experience that most buildings are able to withstand these earthquake forces inelastically. Because of that, arbitrary reduction factors traditionally were adopted for different types of buildings.

Initially, the design force levels in the code were much lower than the forces we knew would be exerted during an earthquake. Engineers could easily calculate elastic behavior of their structures with these reduced force levels. Detailing for ductility and redundancy were then relied upon to allow the structure to withstand the expected higher loads. ATC-3was the first code that recognized the real earthquake forces, and then provided the reduction factors on those loads that bring you back to about where the code was in 1973. ATC-3 introduced specific reduction forces for different structural systems for the first time, but the end result was to provide a design that was about comparable to what you would get with UBC 1973. UBC 1973 did not have the reduction factors, but started with lower earthquake forces and modified them with four coefficients for different structural systems. ATC-3 introduced a more realistic approach because it recognized the real forces.

We have learned since 1972 or 1973 that we can explain some of these things—whereas some others we still cannot explain. The reduction factors in the code today are somewhat arbitrary, and there have been a number of efforts to come up with more realistic reduction factors that are justified by research as well as by actual experience.

Scott: Do you think that the reduction factors are too high, or too low, or a mixture of both?

Nicoletti: I think it is a mixture of both, and this is why many engineers would like to discard the R factors in favor of different approaches, as was done in FEMA 273, the seismic rehabilitation guidelines.

Scott: You must have been going through some pretty complicated discussions on these matters.

Nicoletti: Yes, and it is going to take some time to reconcile various points of view. There have been research projects that are providing background for the discussions. I believe, as more engineers become familiar with the new procedures, there will be more support for them.

BSSC and the Spectral Ordinate Maps

Nicoletti: As the new FEMA provisions are implemented, one thing that is bound to cause some problems relates to recent USGS efforts to define ground motion. USGS has developed and is recommending a series of maps that present ground motion in terms of spectral ordinates, rather than acceleration. The maps have been prepared for various seismic hazard or ground motion levels, such as ten percent probability of exceedance in fifty years (mean recurrence interval of 475 years), and two percent probability of exceedance in fifty years (equivalent to two percent in 250 years, a 2500-year mean recurrence interval). The maps were prepared for BSSC, and were included in the appendix of the 1991 issue of the NEHRP provisions.

Scott: These are the new maps associated with E.V. Leyendecker?

Nicoletti: Yes.

Scott: In lay terms, what is a spectral ordinate map?

Nicoletti: Instead of providing one acceleration value for a given location on the map, these maps provide two points from which the response spectrum can be drawn. A horizontal line is drawn through the first point and a prescribed curve is drawn from the second point to intersect the horizontal line. A different curve is drawn to extend the spectrum beyond the second point. The "spectral" part of the name refers to the x-axis of the graph, which is the spectrum of frequencies or periods; the "ordinate" part refers to the y-axis value, which is the associated acceleration at a given period.

BSSC had decided that they would actually adopt these new USGS maps for the main body of the NEHRP provisions in 1994. In looking at the maps, however, a group of us recognized some real difficulties in their being adopted. They were much too high in some areas and much too low in other areas. We recommended to BSSC that this be investigated, and BSSC set up a Ground Motion Design Values Committee, chaired by Roland Sharpe. I was a member.

To reconcile the matter, we recommended that USGS maps be retained as reference documents, but that design value maps be prepared in which the USGS values have been adjusted—truncated in some cases and raised in others, so the end results would be about the same as we are getting now. Roland met with BSSC and presented our recommendations, but BSSC was not ready to accept them. The controversy did, however, prompt them to put the new maps off until the 1997 edition of the provisions.

Prior to the preparation of the 1997 NEHRP provisions, BSSC convened a new Ground Motion Design Values Committee. It has some of the same members as the 1994 committee, but they replaced a few of the troublesome California members, such as Rol Sharpe and me. The new committee, chaired by R. Joe Hunt, proposed an interesting new approach to address the concerns of engineers in places like Salt Lake City, Utah, New Madrid, Missouri, and Charleston, South Carolina. In recognition of such areas where intraplate seismic activity has long return periods that are not reflected in the ten percent probability of exceedance in fifty years seismic zoning, the committee proposed using the USGS maps with a two percent probability of exceedance in fifty years—about a 2,500 year return period—and modifying the map values to represent what they termed maximum considered earthquake (MCE) values.

The new MCE maps introduced minimum values in very low seismic areas, and truncated high values on the Pacific Coast near large active faults. The proposed design values were established as two-thirds of the MCE values. The intent of this was to provide the same hazard level throughout the U.S. with essentially the same design values in areas of high seismicity, but more realistic values in the central and eastern U.S. that consider the large, but rare, earthquake.

Although USGS was not completely happy about this, the new wrinkle with the MCE maps attracted enough support so that the design value maps were adopted by the 1997 NEHRP provisions. The authors of FEMA 273 and 274 decided to prescribe the 10 percent probability of exceedance in fifty years ground motion for the rehabilitation of existing buildings for the life safety performance objective. Since this is generally less than the two-thirds MCE ground motion, it facilitates rehabilitation. The two FEMA documents prescribe that the rehabilitation shall also comply with a collapse prevention performance level with two percent probability of exceedance in fifty years ground motion. The provisions also stipulate that this latter ground motion need not exceed the MCE values. Thus, the poor engineer designing the rehabilitation of an existing building must refer to three different sets of maps to implement a design.

Chapter 7

Hazardous Buildings

We questioned every one of the provisions, and found that some of them were pretty subjective.

Nicoletti: In recent years, there has been increased interest in existing buildings, and my activity has also shifted in that direction. Even before the Loma Prieta earthquake, there was quite a movement in California to try to mitigate the hazards represented by some of the older existing buildings, particularly unreinforced masonry buildings (URMs).

Senate Bill No. 547

Scott: Those are important issues—would you please spend a little time discussing them?

Nicoletti: The California unreinforced masonry provisions originated with Senate Bill No. 547 (SB 547), a state legislative bill aimed at reducing the seismic risk of so-called potentially hazardous buildings, and were enacted in 1986.[21] Unreinforced masonry buildings were identified as being the most potentially hazardous. SB 547 set up the timetable for communities in California to address their unreinforced masonry buildings. They were required to identify these buildings, and

21. Cal. Gov. Code, §8875 et seq., 1986.

put in place a project or a plan to mitigate the hazard. The bill's language is very loose, however, because it does not require actual implementation of a plan, such as by adopting a retroactive strengthening ordinance. The bill has been interpreted in different ways.

Early Local Efforts

Scott: Long before SB 547 was passed, there were some local efforts at mandating building hazard reduction. I guess Long Beach was the most notable one.

Nicoletti: Yes, Long Beach was the city that initially took the lead in this. Even before the 1971 San Fernando earthquake, they had started addressing hazardous buildings with a program of appraising and condemning or requiring strengthening of pre-1934 URMs, buildings that had been designed prior to the adoption of seismic regulations developed after the 1933 Long Beach earthquake. This prompted a response from property owners, leading to a closer look at the issues. An ordinance proposal was under consideration when the San Fernando earthquake occurred, reinvigorating local concern.

An earthquake hazard ordinance was approved shortly afterward, and the program was strengthened by a new ordinance in 1976. In any event, most of the buildings built before 1935 had already been dealt with when Senate Bill 547 was passed. In 1990, Long Beach enacted an ordinance that picked up the remaining unreinforced buildings. We helped them prepare that ordinance.

Scott: The Long Beach program started before San Fernando, with the local building official, Ed O'Connor, pushing it. Their memory of the 1933 earthquake damage helped. The recent ordinance you just mentioned is a relatively new development to extend the Long Beach program?

Nicoletti: Yes, it was a new ordinance, enacted I think in 1990, and addressed buildings built between 1935 and 1973.[22] I think they adopted the ordinance shortly after the Loma Prieta earthquake.

Los Angeles passed its URM ordinance in 1981 with their amendment to the city building code, which mandated retrofit of the unreinforced masonry bearing wall buildings, as distinct from frame-plus-URM-infill structures.

Soon after the 1971 earthquake, an effort was begun to write some kind of regulations for hazardous buildings in Los Angeles. It took ten years to get the preliminary studies done and the URM ordinance drafted and adopted. Earl Schwartz (now retired), as a member of the Los Angles Department of Building and Safety, played an active role in the development and implementation of that ordinance.

Palo Alto, California took a different approach in its seismic safety ordinance. Adopted in 1986, it identified the potentially hazardous buildings, and notified the building owners.[23] Each owner had to have an engineer make an evaluation of the building, and indicate what would have to be done to bring it up to about the 1973 UBC, which is about three-quarters of

22. City of Long Beach, "Proposed Amendments to Earthquake Hazard Regulations, Title 18, Chapter 18.68, Long Beach Municipal Code." City of Los Angeles, Department of Building and Safety. Adopted by ordinance April 1990.

the current code earthquake force requirements. After submitting this report to Palo Alto and posting it in his building, the building owner had a certain amount of time to notify the city what he intended to do about the building, which could be nothing. According to Palo Alto, that met the intent of Senate Bill 547.

It has not worked too well in Palo Alto, as only a few building owners have actually done the upgrading of their buildings. Most of them have elected to do little or nothing. Some of them have done some of the more pertinent portions of the indicated retrofit, but many of them have decided to do nothing.

Reconciling Several Versions of URM Ordinances

Nicoletti: Because several versions of the URM ordinances had been kicking around, at the end of 1987 the California Seismic Safety Commission asked SEAOC to prepare a consensus ordinance similar to the Los Angeles ordinance. The committee set up to do that was chaired by Al Asakura, of the Los Angeles building department. Al Asakura had succeeded Earl Schwartz as head of the Los Angeles earthquake safety division. I was on the SEAOC committee, which had representation from throughout the state.

The committee SEAOC set up was called the Unreinforced Masonry Buildings Task Committee. We started in 1987 with the latest version of Division 88 of the Los Angeles city building code, and used that as a "straw man," you might say. The Los Angeles ordinance was then numbered Division 88, and we took it and the accompanying Rules for General Application (RGAs), which permitted building evaluation and design in accordance with the ABK methodology[24] as an approved alternative to Division 88, as a starting place. The Task Committee decided to draft a set of provisions permitting either a General Procedure—a modified code approach similar to Division 88—or an alternative Special Procedure based on the ABK methodology in the Los Angeles RGA. At the time, Asakura informed us that by 1988, about 90 percent of the 8,000-plus URM buildings in Los Angeles had been retrofitted, and that about 90 percent of the retrofitted buildings were done under the ABK method of the RGA.

Scott: The original Los Angeles city ordinance was passed in 1981, and became Division 88 of the city code. How did you build on or modify the Los Angeles provisions?

Nicoletti: We questioned every one of the provisions, and found that some of them were pretty subjective. They had evolved over the years. Los Angeles had set up a separate group within their building department to deal with

23. California Seismic Safety Commission, "Earthquake Hazard Identification and Voluntary Mitigation: Palo Alto's City Ordinance, 1990." Gives background on the ordinance's development and enactment.

24. "ABK" refers to a special set of retrofit guidelines developed in the joint venture research project by three Los Angeles area engineering firms, Agbabian Associates, S.B. Barnes Associates, and Kariotis & Associates. A series of reports were published, with the summary volume being: ABK Joint Venture, *Methodology for Mitigation of Seismic Hazards in Existing Unreinforced Masonry Buildings*, Topical Report 08, funded by the National Science Foundation, January 1984.

that ordinance. Many decisions were made between the client's engineer and this group, and these decisions were not represented in the ordinance. We found this out in going over the details. Al Asakura would say, "Well, that isn't the way we do it—we do it this way."

Scott: In your review and reworking, what was your main objective—to reshape the Los Angeles ordinance for statewide application?

Nicoletti: Yes. The committee reorganized the provisions and, after many drafts, and many sessions of considering the validity and applicability of each provision, a final draft was developed in October 1989.

Then in December 1989, SEAOC decided to form a new standing committee. It was originally called the Hazardous Buildings Committee, and later called the Existing Buildings Committee. Four regional SEAOC associations appointed two representatives each to the new committee, and I was appointed to chair it. The new committee—together with SEAOC's Code Committee—was charged with final polishing of the URM provisions, which were eventually published by ICBO in the 1991 edition of the Uniform Code for Building Conservation (UCBC).

A More Economical Alternative

Scott: Would you discuss ABK a bit more?

Nicoletti: The research on which the ABK methodology is based was performed for NSF as a joint venture. It was a big program that went on for several years, and had several million dollars in funding. Out of this, Agbabian and Barnes, two of the partners in the joint venture, have remained silent, whereas Kariotis was the one primarily responsible for the Rules for General Application (RGAs) provisions. While Agbabian and Barnes have not spoken out against the provisions, I do not think they totally endorsed them.

There is also still some question as to whether the provisions can be completely justified by the results of the research done by ABK. In fact, there have been several research efforts to review and confirm the performance of brick wall assemblies with wood diaphragms. As might be expected, the results contain a lot of scatter due to differences in workmanship and construction details.

The ABK methodology is a complete departure from the approach taken by other seismic provisions in the code. I think some of it was very strongly brought out by the research done by ABK. Some of it, however, was not quite so clearly brought out, and is still controversial. The ABK methodology made it possible to retrofit brick buildings much more economically than otherwise.

Scott: How does the ABK methodology achieve the cost reduction?

Nicoletti: The ABK methodology, and also what the Los Angeles RGA eventually said, was that it is *not* a good idea to strengthen diaphragms. This is contrary to everything we had ever done before. Instead, they say that the diaphragm can be used to absorb energy. I emphasize that I am talking about wood diaphragms, because the ABK methodology is applicable only to wood diaphragms.

You do not want a strong diaphragm, but a weak one that does not deflect too much. The weaker the diaphragm, the less shear it transfers to the in-plane walls. If a diaphragm deflects too much, they introduce cross-walls—vertical walls that are dampers (like shock absorbers) that absorb energy but are not calculated to take force (although they actually will). The walls control the deflection of the diaphragm, and keep it from deflecting too much.

The out-of-plane walls are considered flexible as compared to the horizontal rigidity of the diaphragm, which acts as a horizontal beam spanning between the in-plane walls. The out-of-plane walls conform to the diaphragm's horizontal deflection, which must be controlled to prevent excessive distortion and distress in the out-of-plane walls.

The two main new ideas introduced by the ABK methodology were, first of all, the fact that the shear transfer to the wall is limited by the strength of the diaphragm, and second, you want to control the deflection of the diaphragm rather than the strength. While there are many other things, those are the two main ones. They made it more economical to retrofit buildings.

If you use the general procedure—the code approach—the first thing you find out is that the wood diaphragms are not adequate. So you put plywood over them to strengthen them and that makes them stiffer, which transfers more load into the diaphragm, and more shear into the walls.

Scott: In other words, you are creating more problems?

Nicoletti: Yes. The provisions of the Special Procedure were based on the observed nonlinear response of URM buildings rather than the elastic response implied by the General Procedure. While subject to considerable uncertainty, it significantly reduced the seismic risk and was more economical than the General Procedure.

San Francisco's Hazardous Buildings Ordinance

Nicoletti: In 1987, even before the Loma Prieta earthquake, in response to California Senate Bill 547, which mandated that cities mitigate or at least document the seismic vulnerabilities of their unreinforced masonry buildings, the City of San Francisco established an ad hoc committee, which I chaired. The committee was asked to draft an unreinforced masonry ordinance. The ordinance that was adopted in 1992 on our recommendation was based on the 1991 UCBC provision, the Uniform Code for Building Conservation, which originated with the Los Angeles URM ordinance, as I've just explained.

Our ad hoc committee had many questions regarding the applicability of the Los Angeles URM ordinance, so we appointed a special task force consisting of Ted Zsutty, Charles Thiel, and Ron Mayes to review the ABK methodology permitted in Los Angeles. They were able to confirm some of the provisions, but some of them were considered to be questionable. Shortly after the appointment of the ad hoc committee, I was asked to go to Los Angeles to spend a day with Earl Schwartz of the City of Los Angeles Building Department and John Kariotis in order to find out more about the background of Division 88 and the Rules for General Application (RGAs). As a result of the conclusions of our special task force, and the information I got in Los Angeles, we decided that we were going to recommend something similar to the City of San Francisco.

The main thing I learned from the trip to Los Angeles was that the objective of their ordinance was not the same as that of our building code provisions. The objective of their ordinance was called *hazard reduction*, not full *mitigation*. The terms may seem similar from a plain English standpoint, but hazard reduction meant a significant reduction in the risk of injury, which is not the higher level of safety or property protection that was associated with new building design nor with most rehabilitation or risk mitigation projects up to then. The preface to the URM provisions in what ended up in the UCBC, sets out very clearly that the provisions are not intended to *prevent* loss of life or property damage, but to *mitigate* that risk. It was on that basis that we went ahead, both with the San Francisco committee and the statewide committee.

We in San Francisco were looking at Division 88 and the RGAs of the Los Angeles code, and struggling with that. A year later, in 1988, the Seismic Safety Commission asked SEAOC, the state association, to put together the consensus group, which I mentioned earlier, to draft unreinforced masonry provisions that could be adopted as standards by the Commission. Fred Willsea and I were part of that group—there were two of us from each of the four SEAOC associations.

Our ad hoc committee recommended to San Francisco that the provisions of their proposed ordinance be put on hold until the SEAOC consensus group could draft the provisions for the Seismic Safety Commission, as the SSC had requested. We completed these provisions just before the Loma Prieta earthquake, and they were reviewed by SEAOC and ICBO. ICBO eventually adopted the provisions, which are published in the companion volume to the UBC on existing building upgrading. Subsequently, the California Legislature adopted these as the standards for retrofitting unreinforced masonry buildings.

Our ad hoc committee drafted an ordinance for the City of San Francisco based on the UCBC provisions. The City selected a consultant team that included the firm of Rutherford & Chekene to do the environmental impact statement required before issuing the ordinance. The environmental impact statement addressed all the issues—social, economic, and environmental.

One finding was that enforcing this URM ordinance in San Francisco would cause a very severe social-economic impact, particularly affecting residential buildings. The old URM residential buildings were probably the last low-income, low-cost residential units left in San Francisco. The total number of URM buildings in San Francisco that would be affected was something like 2,000, about twenty percent of which were residential buildings.

In another finding, Rutherford & Chekene quantified probable benefits, which were compared with the money spent.[25] Their study showed the greatest benefit for the retrofit money spent would come from anchoring unreinforced walls against out-of-plane failure—using the so-called tension anchors to

25. William Holmes et al. "Seismic Retrofitting Alternatives for San Francisco's Unreinforced Masonry Buildings." A study for the San Francisco City Planning Department by Rutherford & Chekene, San Francisco, California, 1990.

secure the walls—rather than doing a complete building retrofit.

Of course, property owners seized on this. We were asked to modify the ordinance provisions so that providing only the tension and shear bolt retrofitting would still qualify residential buildings as retrofitted. We as structural engineers, through SEAONC, officially opposed this, although we could see the need for it, and the logic of it because the buildings in question represent probably the only remaining low-cost housing in the city.

In 1991, San Francisco decided to reactivate the same San Francisco committee that I had chaired, appointing a Seismic Investigation and Hazard Advisory Committee (SIHAC) to advise the chief administrative officer and the Board of Supervisors. We worked with Larry Litchfield, who headed the building department, and with his representative Pervez Patel. In the planning department, we worked with David Prowler. The committee consisted of engineers, builders, officials, and members of the Planning Commission. I was one of the three members appointed by SEAONC. SIHAC agreed with the recommendations of the ad hoc committee, but it became obvious that what the city wanted was the reduced requirements.

Realizing that San Francisco was determined to adopt the reduced provisions, the ad hoc committee made some recommendations as to conditions for buildings to qualify for this reduced-level of retrofitting. We recommended that buildings not qualify if they were irregular in their configurations or had other structural deficiencies that might make them vulnerable even before the bolts would come into action. The city adopted these provisions, and we recommended the ordinance on that basis.

Next, however, the commercial building owners asked for an escape clause similar to the one for residential buildings. They argued that if residential owners can do that, then other owners should be able to as well. The city agreed, so the San Francisco ordinance permits the "bolts-only" retrofit for all buildings that can meet the requirements. The final version of the ordinance, adopted in 1992, dealt with both tension bolts for the out-of-plane forces and shear bolts for the in-plane forces. It also corrects some of the deficiencies that we identified.

To sum up, the San Francisco committee reviewed the URM ordinance in its various drafts, and generally agreed with the position of SEAONC. We eventually approved "bolts-plus" for residential buildings, but recommended rejection of the final form of the ordinance, which would have permitted "bolts plus" for *all* qualifying buildings. ("Bolts-plus" is the term adopted to include both requirements for shear and tension bolts, as well as correction of defined structural irregularities.) The ordinance was finally approved, despite objections from SEAOC and our committee.[26]

San Francisco and Los Angeles Ordinances Compared

Scott: Could you relate the San Francisco ordinance and these issues to the hazardous buildings ordinance of the City of Los Angeles? I believe Los Angeles allows wall anchors

26. Ordinance No. 225-92, "Earthquake Hazard Reduction." San Francisco Board of Supervisors, San Francisco, California. July 14, 1992.

as an interim measure that postpones further retrofitting for a time. Later, however, the complete retrofitting has to be done.

Nicoletti: The City of Los Angeles permitted a "bolts-only" measure to buy the owner some time, although the complete retrofit had to be completed within a specified period. In San Francisco, however, "bolts-plus" is considered a *permanent* retrofit, and the owner does not have to do any more.

Existing Buildings

Ironically, because Los Angeles had been a leader, and had already been doing retrofitting for a few years, whereas we in northern California had not, we had the opportunity to take a fresh look.

Scott: Say a little about the SEAOC Committee on Hazardous Buildings and its work.

Nicoletti: Now called the Existing Buildings Committee, it was set up in 1988. I am now no longer on the committee, as the membership rotates. I was chairman for the first two years, and past chairman for the third year. The committee has addressed hazardous buildings other than unreinforced masonry, such as steel and concrete frames that are infilled with unreinforced masonry. It has also looked at nonductile concrete frames and flat slab buildings where the flat slabs are used as frames, as well as other hazardous systems. By hazardous systems, I mean those not recognized or permitted by the current code.

Differences Between Northern and Southern California

Nicoletti: The Existing Buildings Committee is an interesting committee in that it brings together different points of view—professional differences in the way structural engineering is addressed, especially as it relates to existing buildings in Los Angeles and in San Francisco. It has been interesting to see these things put on the table and discussed. Sometimes there is agreement, and sometimes there isn't. Moreover, some on both sides still do not agree with the consensus the committee has reached. Obviously, some of these matters are quite controversial.

I believe that what dictates some of the differences is the fact that retrofitting has been pursued more actively in southern California, I think because the City of Los Angeles has been pushing it. In contrast, up here, San Francisco has been dragging its feet.

Anyway, some of the southern California engineers who have already done retrofitting would not like to see something more stringent adopted—that might be interpreted as suggesting that what they had previously done for their clients was inadequate. In reality, however, the remainder of the state simply has had more time to consider alternatives and reflect with hindsight on the earlier efforts in Los Angeles. Naturally, there are strong motivations to capitalize on what has been learned and take advantage of this in drafting any new retrofit provisions.

Scott: So the principal source of disagreement is that engineers in southern California have an investment in the existing retrofitting standards and methodology, and do not want to see those standards called into question?

Nicoletti: Yes, pretty much. I think we were able to reconcile most of it. It was too bad, because it held up some of these provisions. Ironically, because Los Angeles had been a leader, and had already been doing retrofitting for a few years, whereas we in northern California had not, we had the opportunity to take a fresh look.

Writing Compromise Retrofit Provisions

Scott: You were trying to find ways to rethink retrofitting, but without upsetting these southern critics too much?

Nicoletti: Yes. Unfortunately this is the way most codes were written—most codes are compromises. You try to make compromises that won't do violence to the original intent too much, but there are compromises. Consequently, I think engineers should view the code only as a minimum requirement.

I have been working with codes for many years, and it is interesting to see all the objections to code changes. Some of them come from vested interests of engineers, and of course the materials suppliers also have a large vested interest. There is jockeying for position between various materials groups that are competing—concrete and steel for instance. The concrete people don't want to see anything in the code that would give steel an advantage, and the steel people feel the same way about concrete. Sometimes the engineer has to be like Solomon, slicing the baby in half.

In the prior review process—before the IBC—SEAOC Seismology Committee usually prepared a recommendation that went to ICBO for inclusion in the UBC for new buildings or the UCBC for existing buildings. Before it was put into the building code, however, ICBO went through its own process. The materials people belonged to ICBO, and so did some of the engineering societies, and some of the building officials. If there was an objection, and if they could not get the required number of votes for approval, the proposal then went back to the SEAOC Seismology Committee, or was disapproved. In short, proposed changes needed to be compromised enough to pass. Compromises took place either on the ICBO floor, or when the proposal was sent back to the Seismology Committee for reworking.

Scott: What will happen now with the IBC?

Nicoletti: I believe that the process is somewhat similar, except that SEAOC's status has been preempted by the BSSC group that updates the NEHRP provisions. I understand that SEAOC intends to keep publishing the Blue Book and, since many of the NEHRP team are also SEAOC members, perhaps the difference in philosophy will not be significant. One significant difference is that, although the BSSC group will continue to propose updates to the provisions, the actual provisions will now be published in ASCE 7,[27] together with all other design loads (dead, live, wind, snow, etc.).

27. "Minimum Design Loads for Buildings and Other Structures." ASCE Standard 7-02. Structural Engineering Institute, American Society of Civil Engineers, 2002.

Issues in Retrofit Ordinances

Scott: Can you again talk in nonengineering terms a little more about some of the issues before these various committees? I guess much of it has to do with existing buildings, with figuring out how hazardous various ones are, and what might be done about them.

Nicoletti: Well, the City of Long Beach took a very simplistic approach, and some other cities have done something similar. Long Beach said that all buildings built before 1973 are potentially hazardous, and so required the owners to essentially prove that they did not have a hazardous building. The year 1973, or the 1973 edition of the Uniform Building Code, was picked as the threshold. In other words, existing buildings are very difficult to strengthen to current code requirements, and to ease the hardship on the property owners, 1973 was picked as a threshold of acceptability for existing buildings. That's basically about seventy-five percent of current code requirements.

Scott: I guess the 1973 UBC also included some of the most significant code improvements that came out of the San Fernando earthquake of 1971.

Nicoletti: Right. The 1973 earthquake code is supposed to be the first edition of the so-called current thinking in earthquake engineering, although substantial changes have been made since, particularly in 1988. There has been a lot of activity, but the 1973 UBC still represents about seventy-five percent of the basic force level in the current codes.

Historic Preservation and Retrofitting

Nicoletti: Some thirty years ago, our firm did the engineering for the seismic retrofit of the State Capitol building in Sacramento. I was the principal in charge of that project. I do not think that today we would be allowed to do what we did thirty years ago on that historic building.

Scott: Some significant historic preservation laws and regulations have been put in place since then. These do not allow as free a hand for retrofitting historical buildings today?

Nicoletti: That is right. Also, in the case of the California State Capitol, we probably did more preservation than would have been thought of twenty years earlier still. Because of its unique status, we went to great lengths to preserve the exterior walls and to salvage and reinstall the wood trim and the floor and wall finishes, and even the architectural plaster. But in true historical preservation, the new interior structural shell would not be permissible. So I don't think we could get by now with what we did then. In fact, if you really listen to some of the historical architects, you almost feel we should not do anything at all. They want a building made stronger, but they want that done without touching the historical part of it.

Some of them even argue that the very fact of a building having suffered structural damage is in itself of historical significance. The damage itself does not make the building, but becomes part of the historic fabric of the building. This is analogous to the historic Liberty Bell and its crack. The Liberty Bell would lose some of its historic significance if the crack were to be repaired.

Resistance to Mandated Retrofitting

Nicoletti: Most people agree with the need for seismic strengthening of public and historic buildings. Nevertheless, as we found in San Francisco, there is a lot of public resistance to retroactive provisions mandating the strengthening of privately owned buildings. This is obviously a delicate matter, and it is very difficult for any jurisdiction to ask a property owner to strengthen his or her building retroactively. The owner has a building that presumably was built in accordance with the regulations in effect at that time. Even though this might have been twenty or thirty years ago or more, the building was approved by the building department at the time.

In short, a building owner has built up a cash flow, or rent income, or whatever—an economic system based on the use of that building. Then, if a lot of money must be spent to strengthen the building in order to continue its use—the alternative being condemnation by the city—the owner suffers a financial hardship. This is one of the biggest drawbacks to mandated retrofitting, and there is no good solution as yet. There is talk about subsidies and ways of making this more palatable to the building owner, but a good way has not yet been found.

NEHRP Seismic Evaluation and Rehabilitation Manuals

Nicoletti: In 1990, FEMA selected our firm to prepare the *NEHRP Handbook for Seismic Rehabilitation of Existing Buildings*.[28] This handbook was intended as one of a series of documents dealing with the evaluation and rehabilitation of existing buildings. The handbook is essentially a menu of available techniques that a designer could look at and say, "I have these deficiencies, and here are the various techniques I might use to upgrade my building."

As a companion document to the handbook, in 1995, we prepared a supplement entitled "Selected Rehabilitation Techniques and Their Costs." It included complete and detailed design examples for six representative buildings and nonstructural components. I was the principal author of the handbook and the supplement. Unfortunately, in spite of our objections, these documents had to be prepared in the absence of any rehabilitation criteria, so the documents had to reference deficiencies and rehabilitation with respect to provisions for new construction. Although the supplement received favorable reviews, it was not published, since FEMA decided to develop similar documents based on the new criteria in FEMA 273.

When FEMA set out to develop guidelines for the rehabilitation of existing buildings, they initially contracted with ATC to identify the critical issues and develop a work plan. I served on the project engineering panel.

The results were published as ATC-28.[29] The report recognized that analytical procedures would be required that could identify deficiencies in structural components that were part of systems that may not comply with current codes. Upon receipt of ATC-28, FEMA contracted with BSSC to develop the guidelines.

BSSC served as the Program Manager, with ATC and the American Society of Civil Engineers as subcontractors. ATC was charged with the task of drafting the guidelines. The ATC guidelines team looked at a number of analytical procedures and finally adopted two procedures similar to what Sig Freeman and I had developed for the Department of Defense manual in 1986. They adopted a linear procedure with *m* values similar to our inelastic demand ratios (IDRs) and a nonlinear procedure with a variation of our capacity spectrum. The big improvement over our Tri-Service documents was in providing acceptance criteria for many more specific structural components based on actual research results. The work by ATC was performed as ATC-33. I served on the panel that developed the concrete provisions. ATC-33 was published only as a working draft for the project team.

The final draft of ATC-33 was used to develop FEMA 273 and 274.[30] These two FEMA documents represented a radical departure from the seismic procedures currently codified at the time. The older UBC provisions prescribed seismic coefficients that represented about one-fourth of the force levels associated with the expected ground motions, and then modified them with four K factors to account for the type of structural system, ranging from 0.67 for moment frame systems to 1.33 for bearing wall

28. FEMA 172, *NEHRP Handbook for Seismic Rehabilitation of Existing Buildings*, 1992.

29. ATC-28, *Development of Guidelines for Seismic Rehabilitation of Buildings*, Phase 1: Issues Identification and Resolution, 1992.

30. FEMA 273, *NEHRP Guidelines for the Seismic Rehabilitation of Buildings*; also FEMA 274, *NEHRP Commentary on the Guidelines for the Seismic Rehabilitation of Buildings*, 1997.

systems, with all of the resulting design forces used in conjunction with allowable stresses.

ATC-3 and the subsequent FEMA/NEHRP documents specified force levels associated with realistic ground motions, modified with a lengthy table of R values for various structural systems, at specified ultimate strength values. Note that these force reduction factors, R values, apply only to the seismic forces. The R values are global reduction factors, factors that apply to the overall building, that assume that each specific system, if it is designed for force levels lower than the elastic forces associated with the expected ground motions, has adequate ductility and integrity because of other design provisions.

In contrast, however, FEMA 273 and 274 recognized that any force reductions based on ductility should be applied to the individual structural components rather than the global structural system. The two manuals that we had prepared for the military also recognized that. In addition, as I indicated earlier, FEMA also adopted modified versions of the linear and nonlinear analytical procedures we had developed in our Tri-Service Manuals.

There seems to be an increasing tendency with model codes to incorporate technical provisions by reference. For example, FEMA 302 incorporates provisions from AISC and ACI and thereby shortens some of the material sections to a few pages of exceptions or additions. The council that publishes the International Building Code has indicated that the FEMA 302 seismic provisions that they have incorporated in the IBC are too lengthy and should be standardized so that they could be incorporated by reference. In response to this expressed need for seismic standards, FEMA has initiated a program to convert some of the newer provisions into standards.

FEMA 178 was the specified reference by the Interagency Committee for Seismic Safety in Construction (ICSSC) for the seismic evaluation of federally owned or leased buildings in response to Executive Order 12941. The FEMA 178 procedures were applied to numerous buildings by many engineers. The consensus was that, while the procedures were generally effective and efficient, a number of improvements could be made. Additionally, FEMA 273 introduced many new analytical approaches tailored to existing buildings. FEMA decided not to update FEMA 178, but to publish a new document and make it a standard that could be incorporated by reference.

The American Society of Civil Engineers (ASCE) was selected to produce the new document. ASCE was also the sponsor for ASCE 7,[31] which is incorporated by reference in FEMA 302. ASCE selected a project team, chaired by Mel Green, and a steering committee that I chaired. The document was published as a prestandard in 1998, has been subjected to balloting by a voluntary standards committee, and later was subjected to public ballot.

FEMA has also contracted with ASCE to publish FEMA 273 as a standard that is currently designated as FEMA 356, but eventually will have an ASCE number. Following the initial publication of FEMA 273 in 1997, BSSC selected about forty existing buildings through-

31. ASCE-7. "Minimum Design Loads for Buildings and Other Structures." ASCE Standard 7-95, 1995.

out the U.S. and contracted with a number of engineering firms to subject them to the FEMA 273 provisions. URS was awarded one of the test buildings, and I was involved in the review of the work. Our conclusions were similar to those of many other firms: that FEMA 273 is a rational and innovative document, but there are many things that need clarification and/or improvement. I participated in the balloting as a member of the standards committee on the first and second draft of the document as a standard. The third draft, with a number of changes, was issued for public ballot in the spring of 2001.

Personal Reservations Regarding Seismic Standards

Scott: With FEMA support and urging, there has been a lot of this kind of work on seismic standards.

Nicoletti: Yes. I have serious concerns regarding the current trend towards creating standards for seismic design, evaluation, or rehabilitation. Experience has shown that the relatively short three-year cycle for building codes is often not able to keep up with the rapid changes in seismic provisions, and current code provisions are far more flexible than an established national standard. Standards are required to conform to the requirements of an approved organization such as the American National Standards Institute (ANSI) and, judging by what is happening with FEMA 310 and 273/356, this process can take from five to seven years. I can agree that some of the procedures—representation of ground motion, equivalent lateral force analysis, dynamic analyses, nonlinear analyses, etc.—could be standardized, but I believe that the detailed provisions should be in a form that can be readily revised or amended.

Impact of Loma Prieta and Northridge Earthquakes

Nicoletti: Following the Loma Prieta earthquake of 1989 and the Northridge earthquake of 1995, I was involved in a number of projects associated with the seismic evaluation and rehabilitation of buildings that had been damaged by an earthquake or were considered to be vulnerable to a future earthquake.

Telephone Buildings

Nicoletti: Two adjacent fifteen-story telephone equipment buildings in the city of Oakland were damaged by the Loma Prieta earthquake. EQE Engineers was hired by PacBell to retrofit the buildings, and I was appointed to chair a three-person advisory panel consisting of Steve Johnston, Professor Steve Mahin, and myself. Following the Loma Prieta earthquake, a special panel of seismologists convened by USGS published a report that predicted a magnitude 7 earthquake on the nearby Hayward fault with a 67 percent probability in the next thirty years. Our advisory panel recommended that the rehabilitation should be designed to resist the predicted ground motion with damage that would permit continued occupancy and equipment function. I suggested using the IDR methodology from our Tri-Service Manual. The procedure would be to select appropriate IDRs for the various structural components when subjected to the predicted ground motion. The buildings were initially subjected to the ground motion in their pre-earthquake undamaged condition so that deficient components could be identified. The components were then strengthened and re-analyzed until they met the target inelastic demand ratio (IDR). A testing program was developed to establish the appropriate IDRs for the wall elements before and after retrofitting.

Coincidentally, the City of Oakland enacted an ordinance that required all buildings that had sustained a loss of more than 10 percent of their pre-earthquake seismic capacity to be strengthened to comply with the 1988 UBC force levels. This was interesting because, at the time of the Loma Prieta earthquake, the City of Oakland was only enforcing the 1976 UBC seismic provisions. The strengthening criteria that we had developed were very close to the requirements of the Oakland ordinance, provided that the buildings were classified as being "essential." Since they contained the critical switching capability for 911 calls in the Bay area, as well as most of the public and private telephone circuits in the east Bay, we concluded that our criteria did not exceed the requirements of the ordinance, and PacBell was able to recoup most of the repair and retrofit costs under the terms of its earthquake insurance policy.

San Francisco Customs House, Civic Auditorium, City Hall

Scott: Have you had the opportunity of working on any other historic buildings since your work on the California State Capitol?

Nicoletti: Although I personally had very little to do with the actual retrofitting design, I provided review and guidance for the recent retrofit of two historic buildings, the U.S. Customs House and the Civic Auditorium, both in San Francisco. The retrofit design was performed in our office and I was delighted to find that the original structural drawings for the Civic Auditorium were signed by Charles Derleth, who was the dean of engineering and one of my professors at Berkeley, as I described earlier.

The firm of Forell/Elsesser was responsible for the design of the seismic retrofit of the San Francisco City Hall and proposed base isolation to minimize damage to the exterior masonry and stonework as well as the interior finishes. Notably, however, even with base isolation, the building and the dome required considerable strengthening to protect the fragile historical materials and finishes.

I was involved in the peer review process for City Hall and also in the value engineering. Value engineering is the process by which a qualified reviewer, who has not been involved in the design, reviews the entire project objectively, and without any imposed constraints, and proposes alternative solutions to achieve the project's objectives. The owner and other interested parties can then compare possible cost savings with other tradeoffs involving, for example, aesthetic or historical significance. In this project, some of the value engineering suggestions were accepted and implemented, while others, some involving larger cost savings, were rejected because of their impact on the building's historical fabric.

Proposition 122

Nicoletti: Proposition 122 was a bond issue of $300 billion for retrofitting state-owned buildings and other public buildings. Fred Turner, of the Seismic Safety Commission, was very active in this program. The first task was to set the guidelines and scope of the research funded by Proposition 122. A small percentage of the $300 million bond issue was set aside for planning and research, primarily to establish guidelines and criteria for the retrofit work. This was pursued in connection with the state Existing Buildings Committee, and with SEAOC and ATC, and several meetings were held.

Chuck Thiel first set forth a general guideline, and then another group headed by Eric Elsesser and Chris Arnold went into more detail, actually developing a plan. They got suggestions and recommendations from a workshop on implementing their plan. They tried to identify how much funding would be needed for drafting criteria, how much for research, and how much for estimating—for cost-benefit studies.

I was marginally involved in this program through ATC-37 and as part of an oversight panel for the review of analytical provisions for the evaluation of existing concrete buildings. Both projects were funded and published by the California Seismic Safety Commission. ATC-37 was published as Report No. SSC 94-03.[32] The other document was published as Report No. SSC 94-01.[33]

Scott: As you may recall, I, too, was a member of the oversight committee.

32. Jack C. Moehle, Joseph P. Nicoletti, and Dawn E. Lehman, *Review of Seismic Research Results on Existing Buildings*. Seismic Safety Commission Report No. SSC 94-03, 1994 (ATC-37).

33. *Seismic Evaluation and Retrofit of Concrete Buildings*. Seismic Safety Commission Report No. SSC 94-01, Volumes 1 and 2, November 1996.

Nicoletti: Yes. I recall that we had some pretty heated discussions. URS subsequently participated in the retrofit of several state-owned buildings under this program, but I was not directly involved.

Welded Steel Frame Buildings

Nicoletti: The Northridge earthquake of 1994 produced some of the strongest ground motion ever recorded and caused so much damage that some accepted design provisions were questioned and reviewed. Probably the most surprising outcome was the extensive damage in welded steel frame buildings. While there were no collapses of such buildings, the nature of the weld failures has raised serious concerns about the safety of all modern high-rise buildings with this type of beam-to-column joint in highly seismic regions throughout the world.

Scott: What is being done with respect to new construction of steel moment frames?

Nicoletti: It is comparatively easy to address the problem for structures not yet built, and elaborate research and testing projects are being undertaken to do so. The problems are enormous, however, in trying to deal with existing welded steel structures, as there are great difficulties in exposing welded joints, evaluating them, and retrofitting those that need it. A joint venture of SEAOC and ATC, which I've mentioned before, and CUREE, Consortium of Universities for Research in Earthquake Engineering, was formed and designated as the SAC Steel Project. I've also served on the Board of CUREE, in the days when it was called California Universities for

Research in Earthquake Engineering. SAC was formed to seek public funding and develop a rational and consensual solution to these welding problems. The work was funded by FEMA and the California Office of Emergency Services. I was appointed to the oversight panel for the first phase that addressed the repair of damaged joints and defined the research and other efforts of the subsequent phases. SAC has published several reports with recommendations for repairs and new construction. A final report was issued in the year 2000.

Chapter 9

Caltrans and Peer Review

The retrofits were not adequate to ensure that the freeways would not be damaged in a similar way, in a repeat of the Loma Prieta earthquake.

Nicoletti: Shortly after the Loma Prieta earthquake in 1989, Caltrans responded to recommendations by the Governor's ad hoc Board of Inquiry by beginning to put together peer review panels to review the repair and retrofit designs.[34] Several separate Caltrans review panels were set up. The first was the San Francisco viaduct panel, which I co-chaired with Nick Forell, the Terminal Separations project—the redesign and relocation of the Cypress viaduct in Oakland, and another a panel for the 980-24-580 interchange in Oakland.

Scott: You were on four peer review panels for Caltrans, weren't you?

---

34. Housner, G. W., et al., *Competing Against Time: Report of the Governor's Board of Inquiry on the 1989 Loma Prieta Earthquake.* State of California, Governor's Office of Planning and Research, 1990.

Nicoletti: Yes. The funding for the seismic strengthening of the toll bridges is a separate appropriation, and later I will discuss my involvement with the east crossing of the San Francisco Bay Bridge.

Six San Francisco double-deck viaducts were closed following the Loma Prieta earthquake. I was on the peer review panel Caltrans appointed for those repairs. There were six engineers on the panel, plus four academic advisors from the University of California at Berkeley and at San Diego. Caltrans selected six consultants to retrofit the six freeways. Subsequently, San Francisco decided to demolish the Embarcadero Freeway and Caltrans decided to replace the Terminal Separation structures with new construction.

The remaining San Francisco viaducts to be retrofitted, the elevated freeways, included the Central Freeway, the Interstate 280 (I-280) interchanges at U.S.-101 and Alemany, the remainder of I-280 to Army Street, and the end portion of I-280 at China Basin. Subsequently, the Central Freeway went through some political maneuvers, and only a portion of it was demolished and rebuilt. The retrofit design for the I-280 viaducts was the only one that was carried through completely to construction, and was completed in 1995.

While Caltrans intends to design most new construction in-house, the freeway repairs in San Francisco were put out to consultants, so that was a different type of peer review. It is essentially a different function and works on a different basis, because the consultants were given instructions by Caltrans, and we were a third party. Whereas when Caltrans is doing the work, it is a more direct relationship.

San Francisco Double-Deck Viaducts

Nicoletti: Caltrans reacted very quickly, perhaps too quickly, to the need to retrofit the San Francisco freeways. First, they selected separate consultants and gave them whatever guidance they could at the time. By the time our review panel was formed, the consultants had already done some preliminary designs, and in fact some construction contracts had been awarded on the basis of these preliminary designs.

It very soon became obvious that these retrofits would not satisfy the requirements of the Governor's Board of Inquiry. The retrofits were not adequate to ensure that the freeways would not be damaged in a similar way in a repeat of the Loma Prieta earthquake. So a big decision was made, and we recommended that they terminate all work and start over again.

Caltrans eventually agreed, and all the construction and design work was terminated. New criteria were developed and implemented for the retrofit. In the meantime, a few of the repairs that had been made were left in place. You can still see some of the bolted steel plates on the Mariposa Street onramp to I-280. These may be removed eventually, but have been left in place for the time being as they do not impact the primary structural support of the freeway. Most of the preliminary repairs had to be removed and redesigned, as it was obvious that the deficiencies were not only in the columns, but also in the column-to-beam joints.

There was a design deficiency that had to be corrected, although it is hard to call it a "deficiency," because that was pretty much state-of-the-art at the time of design. Now, however, the joints are recognized as vulnerable. So the recommendations were that some of the columns and the joints be replaced. Essentially, the deck had to be supported while the columns and the column-beam joints were being replaced. In many cases, an external girder was added on each side of the freeway at the lower level to take the longitudinal forces. The freeways look somewhat the same, but will be a lot stronger.

I don't know if anyone had made an estimate of the cost of wholly new construction. But there were many reasons for retrofitting, as opposed to new construction. The main reason had to do with maintaining a certain amount of the vehicle traffic flow capacity. Some capacity had to be kept in service while the work was being done. Retrofitting lent itself to continued operation of the bridge much better than building a wholly new structure would have.

The design for the retrofit of the San Francisco viaducts was completed and the panel submitted its final report in 1994. The panel was not officially disbanded, however, because of possible changes in the retrofit of the Central Freeway and the U.S.-101 and Alemany interchanges, but as it turned out, no further action by the panel was required.

Bay Bridge On-Ramps and Off-Ramps

Nicoletti: A second freeway project, called the Terminal Separations Project, had to do with the on-ramps and off-ramps for the Bay Bridge. That was originally scheduled for retrofit by Bechtel, but when the preliminary plans were completed, it was obvious that the retrofit would be very complex. It would be more cost-effective to demolish and rebuild the freeway. As a matter of fact, some traffic engineering decisions changed, and they, in turn, implied some changes in structural decisions, and it made more sense to demolish and rebuild rather than retrofit.

Caltrans elected to do this in-house in Sacramento, but they put together a panel (which was actually a portion of our San Francisco panel) to do the peer review. I was chairman of that panel, which reviewed the work that the Caltrans people were doing. The design for this project was completed, but construction was held up because San Francisco decided to demolish and abandon some of the ramps that were scheduled for replacement.

Other Panels

Scott: You chaired the panel for the Bay Bridge approaches. But there were also separate panels for other San Francisco freeways, weren't there?

Nicoletti: The other three San Francisco projects—for I-280, the Embarcadero, and the Central Freeway—were reviewed by the first panel that I co-chaired. A separate three-member peer review panel was convened for the redesign of the Cypress Street viaduct in Oakland. I was on that panel, which was chaired by Professor Frieder Seible of U.C. San Diego. We completed our work in 1992, but the project was held up for some time by the residents adjacent to the new alignment of the freeway. Eventually, however, the alignment issue was resolved, and construction was completed in 1999.

I was also on a fourth panel, chaired by Professor Jack Moehle of U.C. Berkeley, that reviewed the evaluation and retrofit of the 980/24/580 interchange, the "MacArthur Maze." The reconstruction of that major interchange was completed in 1999.

Peer Review

Scott: Would you discuss how a peer review operates?

Nicoletti: The panels meet periodically, initially about once a week, or every other week. The consultants make a presentation on what they are doing, the criteria they are following, and the design problems they are addressing. We critique their approach to the problems, and then make suggestions, or sometimes recommend different approaches. Many of these problems are unique. Something of this magnitude has not been done before, and many of the design procedures are state of the art.

Scott: Discuss the review process and how it worked.

Nicoletti: I believe it was a learning process for everybody. I think everybody accepted it in the spirit in which it was intended. We had the academic people, who gave us the principles. We had the practical engineers, who could resolve some of the practical issues. Caltrans was very cooperative and supportive, and the process went right along. I think it resulted in a more rational design for the retrofitted structures, and there is also a beneficial influence on new structures designed by Caltrans.

Caltrans refers to the panels as peer review panels, but in the way they are formulated and act, they are not truly peer review panels. To me, a peer-review means one made up of people who have about the same qualifications as the ones who did the work being reviewed. These panels, however are made up of experienced practicing engineers and professors from universities. The purpose of a conventional peer review is to establish the "state of the practice." In contrast, I think these panels are intended to establish the "state of the art." They are proposing different and innovative approaches to the analyses and design.

Scott: Discuss the difference between "state of the art" and "state of the practice."

Nicoletti: Let me start with the "state of the practice." Normally, peer-review panels are set up to determine whether the work has been done according to the state of the practice. Have the codes been followed correctly? Are proper and appropriate procedures used? Whereas a "state of the art" review would look to see whether the state of the art has been used, and may involve a quite different approach to analysis and design than is currently being done by the professional field in general.

Scott: "State of the practice" means what is generally accepted at the time by the good engineers in the profession. "State of the art" maybe goes on up further and beyond what is generally practiced at the time. Are those fair statements?

Nicoletti: Yes. Given the fact that there was essentially no precedent for retrofitting major bridge structures following a severe earthquake, the state of the practice in this case was almost irrelevant.

Scott: Ductility is an important consideration in buildings. Is it also an important consideration in bridges?

Nicoletti: In bridges, the primary lateral force-resisting components are the columns, and detailing for ductility is very important. In other structural systems, perhaps some of these considerations could be overlooked because of redundancy. In a very simple major structure that has very little redundancy—like an elevated freeway—ductility is very important, as are the details intended to provide ductility.

Scott: So relative lack of redundancy makes ductility even more important?

Nicoletti: Yes. Bridge structures are simple, and have little redundancy. In contrast, a building has many columns, walls, partitions, and a lot of redundancy. Thus, a building has a lot of alternate paths for forces to be resisted. In a bridge with four columns and a deck, you don't have those alternate paths.

Ductility basically means the ability to withstand displacements greater than the elastic capacity of the structure. When there is no backup, ductility and the details to provide it are very important. If the details don't do the job in a nonredundant structure, there is no other backup.

Caltrans Seismic Advisory Board

Scott: In addition to the review panels, you are also on the Caltrans Seismic Advisory Board, aren't you?

Nicoletti: Yes. In response to the Governor George Deukmejian proclamation following the Loma Prieta earthquake, Caltrans established a permanent Seismic Advisory Board as a continuing body to review all of their criteria for new construction, as well as existing structures. I am on that board, which was initially chaired by Professor George Housner, who also chaired the Governor's ad hoc Board of Inquiry. Professor Housner retired from the Seismic Advisory Board in 1998. Professor Joseph Penzien served as the chair until his resignation in 2004, and Professor Frieder Seible is the current chair.

Immediately after the Northridge earthquake of January 17, 1994, the Seismic Advisory Board inspected the damage to highway bridges in the affected area. Our report not only discussed the bridge damage, but also evaluated the response by Caltrans to the Governor's proclamation following the 1989 Loma Prieta earthquake.[35]

The Seismic Advisory Board meets quarterly. Its function is not, however, to review designs. Its purpose is to review and make recommendations regarding policy to be adopted by Caltrans with respect to seismic design criteria and procedures. It reviews current design criteria and design and analysis procedures, and makes recommendations to the state Director of Transportation. Again, Caltrans has accepted this very well.

Caltrans-ATC Contract (ATC-32)

Nicoletti: In late 1991, Caltrans funded an ATC project, ATC-32, to review all current criteria in Caltrans design manuals and propose additions and revisions to bring them up to date. The work was done under the traditional ATC format—a thirteen-member project engineering panel was appointed. I was

35. George W. Housner et al., *The Continuing Challenge: The Northridge Earthquake of January 17, 1994*. Report to the Director by the Caltrans Seismic Advisory Board, California Department of Transportation, 1994.

one of the panel members. Individual contractors were selected to develop specific tasks, and the results were reviewed and critiqued by the panel. The end product was a report, *Improved Seismic Design Criteria for California Bridges*,[36] with recommended revisions to the Caltrans documents.

Scott: I take it that this project, and the other ATC projects, focused mostly on seismic design concerns?

Nicoletti: Seismic concerns were uppermost, but ATC-32 also incorporated some of the things that came out of the peer review process for the elevated viaducts. These were state-of-the-art procedures for dealing with ductility, analysis, and design. Caltrans has adopted many of the ATC-32 recommendations. In addition, based on the subsequent experience with the toll bridge retrofits, they modified some of the ATC-32 recommendations prior to adoption in their manuals.

Scott: Is there a published version of ATC-32 available to the public?

Nicoletti: I am not sure. All publications for public agencies are in the public domain, however, and should be available from the specific agency.

A Great Opportunity for Engineers and Caltrans

Nicoletti: Caltrans was mandated by the Governor to follow the ATC-32 recommendations. They were also required to make periodic reports on what they are doing, and they have substantially implemented the ATC recommendations with a few modifications.

Scott: This whole thing put some very important processes in motion, didn't it?

Nicoletti: It did. And it was also a great opportunity for the engineering community, as well as for Caltrans. These opportunities don't come very often. They are opportunities to take a look at what has been done, in view of what you know now, and to make appropriate revisions. Usually the codes are patched in bits and pieces.

It was also a way of getting Caltrans to be less insulated from the rest of the discipline. Here I am giving my impression, based in part on impressions I have picked up from talking to engineers and nonengineers. In the past, Caltrans was perhaps a little too insulated from the rest of the structural engineering community. These developments have helped break down that insulation and improve the situation.

Scott: Caltrans appears to have had good direction from the top in trying to deal with the aftermath of Loma Prieta, at least in trying to do things as nearly right as they could, although obviously they had to learn as they went along.

Nicoletti: I think that is true, and I think a lot of it is due to Jim Roberts, the recently retired head of engineering services. I think he was very dedicated to learning from the past and building for the future and doing it right. Jim is currently a member of the Caltrans Seismic Advisory Board.

36. ATC-32, *Improved Seismic Design Criteria for California Bridges: Provisional Recommendations.* Applied Technology Council, Redwood City, California, 1996.

East Crossing of the San Francisco-Oakland Bay Bridge

It was apparent to me that retrofitting the east crossing of the bridge to ensure adequate performance was going to be a very difficult task.

Nicoletti: The Loma Prieta earthquake of 1989 forced the closure of the San Francisco-Oakland Bay Bridge when an upper deck span was displaced and one end fell from its bearing at pier E9, between Yerba Buena Island and the City of Oakland.

The damaged section was located in the easterly portion of the crossing where the bridge consists of double-deck simply-supported truss spans resting on concrete piers, which are in turn founded on relatively shallow timber pile foundations in the deep bay sediments that are as much as 400 feet deep. Pier E9 occurs at a change in horizontal alignment where the bridge turns to the south towards the Oakland toll plaza. The damage occurred because the bearing seats and connections at the upper deck level could not accommodate the large relative displacements in response to the ground motion.

Shortly after the Loma Prieta earthquake, the U.S. Geological Survey convened an ad hoc group of local seismologists to discuss the seismic hazard in the San Francisco Bay area. One of their conclusions was that there was a high probability of a magnitude 7 earthquake on the Hayward fault in the next thirty years. The Loma Prieta earthquake damage to the Bay Bridge was caused by a magnitude 7 earthquake seventy miles away on a segment of the San Andreas fault, whereas the Hayward fault was only three miles from the Oakland end of the bridge.

BCDC's Engineering Criteria Review Board

Scott: I understand that you are involved with the current plan to replace the San Francisco Bay Bridge's east crossing, the part of the bridge between Yerba Buena Island and Oakland.

Nicoletti: Yes, I am, both as a member of the Caltrans Seismic Advisory Board and as past chairman and current member of the Bay Conservation and Development Commission's Engineering Criteria Review Board (ECRB). BCDC has authority over all structures that are on San Francisco Bay.

Scott: Is the ECRB kind of a peer review group?

Nicoletti: No. The Engineering Criteria Review Board is much more than a peer review panel. It is an established board within BCDC. We review the criteria for projects and make recommendations to the Commission. These projects need BCDC approval before they can go to the building department and get a building permit. Many of the structures involved are waterfront structures and are not regulated by the building codes and the ECRB needs to approve the criteria developed by the applicant.[37]

Engineers from Caltrans appeared before the ECRB to describe the damage to the San Francisco-Oakland Bay Bridge from the Loma Prieta earthquake, the probable causes, and the proposed repairs. Caltrans indicated that the proposed repairs were temporary and a complete evaluation and retrofit was scheduled. As a member of the Engineering Criteria Review Board, it was apparent to me that retrofitting the east crossing of the bridge to ensure adequate performance was going to be a very difficult task.

Right now the biggest ECRB projects are the review of the work proposed for the bridges across the Bay. Caltrans has a program for strengthening all of the toll bridges across the Bay, including the major bridges like the Bay Bridge and the San Mateo Bridge, and of course the Golden Gate Bridge District has already completed a program for strengthening the

37. BCDC established the twelve-member Engineering Criteria Review Board as part of the process of implementing its comprehensive plan for the governance of San Francisco Bay. A committee chaired by Karl V. Steinbrugge prepared the background study for BCDC on which the concept was based. *Carrying Out the Bay Plan: The Safety of Fills*, San Francisco Bay Conservation and Development Commission, 1968. The committee members were also appointed as the initial members of the review board.

Golden Gate Bridge. They all have to come to our ECRB for approval of their criteria.

Replacement of Bay Bridge

Nicoletti: The Caltrans Seismic Advisory Board (SAB), of which I am a member, was informed by Caltrans that evaluation and retrofit of the seven California toll bridges were to be funded separately from the approximately 27,000 other state highway bridges. Because all the state bridges were designed in strict accordance with their design manuals, it was fairly easy for Caltrans to identify potentially deficient bridges by the design date and the presence of details that had been found to be vulnerable. A total of about 7,000 bridges were identified as requiring evaluation and possible retrofit.

In 1995, when the retrofit program for conventional bridges was about ninety percent complete, Caltrans began to address the retrofit of the seven toll bridges. The agency contracted with private engineering firms for six of the toll bridges, but they decided to tackle the retrofit of the San Francisco Bay Bridge with their in-house engineering staff.

The Seismic Advisory Board received periodic reports on the progress of the evaluation and retrofit concepts. As the evaluation neared completion, it became apparent to several of us on the Board that Caltrans was not going to meet its performance objectives without major work on the foundations of the Bay Bridge, as well as the fragile superstructure. In addition, the retrofit was approaching the estimated cost of replacement, and would require closing the bridge to traffic for extended periods. In November 1996, in a letter to the Director of Transportation, the Seismic Advisory Board expressed its concern and strongly recommended that, in lieu of retrofit, Caltrans should consider replacement with a parallel bridge. Caltrans agreed and initiated conceptual designs for replacement.

Scott: So the Seismic Advisory Board recommended that the bridge in question be replaced rather than retrofitted?

Nicoletti: Yes. We met with the Peer Review panel for the retrofit and they endorsed our recommendation.

Selecting a New Design

Nicoletti: The Caltrans concept envisioned two parallel concrete viaducts from the Oakland toll plaza to Yerba Buena Island. The Metropolitan Transportation Commission (MTC), with representatives from the nine Bay area counties, decided that they should have the option of selecting the appearance and alignment of the new bridge. Governor Pete Wilson agreed, but with the provision that the selected bridge should cost no more than $200 million more than the Caltrans concept and that the additional cost would be raised by increasing the toll charge to $2.00 per car.

MTC appointed an ad hoc task force, consisting of members of the commission and chaired by Mary King, the representative from Alameda County. The task force recognized the need for technical support, and in February 1997, appointed an Engineering and Design Advisory Panel (EDAP), consisting of existing committees or boards within BCDC and Caltrans and additional members from the local chapters of various professional societies. I was appointed to chair this group of thirty-six outspoken and diverse individuals.

Scott: It sounds like a group that could be very difficult to manage.

Nicoletti: Actually, it wasn't as difficult as it would seem. After I went around the table and gave everyone a chance to speak, it was amazing how readily we reached consensus.

During the course of several meetings, EDAP identified the principal aesthetic and functional design issues and developed guidelines for the design of the crossing. In response to direction from MTC, and dictated by the existing natural deep channel as well as the need for visual impact, a "signature" span was specified adjacent to Yerba Buena Island. EDAP then convened a three-day workshop and invited conceptual presentations for the new crossing in accordance with the established guidelines.

A total of twelve concepts were presented, and by May 1997, EDAP had narrowed the signature span options to either a self-anchored suspension span or a cable-stayed span with the remainder of the crossing to be two parallel concrete viaducts as Caltrans had proposed. EDAP recommended that MTC request Caltrans to have two independent design teams develop each option to thirty percent completion for a final decision. This was done, and, by a narrow margin, the self-anchored suspension span was the preferred option. The joint venture of T.Y. Lin International and Moffat and Nichol was selected by Caltrans for the final design. Caltrans also selected a peer review panel, and I was selected to chair the panel, whose other members are Frieder Seible, I.M. Idriss, Ben C. Gerwick, and Jerry Fox.

Construction Begins—More Problems

Nicoletti: Although the deliberations of EDAP had been conducted in a public forum, no serious objections were heard until the final design was well underway. At this point, in spite of the fact that both cities had tacitly approved the EDAP selection, the City of San Francisco (Mayor Willie Brown) wanted to move the new bridge to the south of the existing bridge. The City of Oakland's mayor, Jerry Brown, wanted an entirely new design. Fortunately, Governor Gray Davis and the State Legislature supported the selected design, but San Francisco was able to establish several obstacles that have delayed the bridge by almost two years.

Scott: You mean construction of the new replacement bridge may soon actually begin? I suppose one could say "It's time!" —the Loma Prieta earthquake was about twelve years ago.

Nicoletti: [Portions of Nicoletti's comments here were added after the last 2001 interview sessions with Scott, bringing the account up to date as of spring, 2005.] The construction is divided into a number of contracts; a few are small enough to allow local contractors to compete. The four major contracts that require large contractors or joint ventures are: a) the concrete viaduct or Skyway, b) the self-anchored suspension (SAS) bridge, c) the Yerba Buena Island transition, and d) the Temporary Bypass. The Skyway contract was advertised in 2001 and only one bid, significantly in excess of the budget, was submitted by a joint venture headed by Peter Kiewit Construction. After some deliberation, the contract was awarded and construction of the Skyway is currently (January 2005) on schedule and about sixty percent complete.

The SAS contract was advertised in 2002, but the bid opening was delayed twice and finally opened in September of 2004. Governor Davis had requested federal funding for this program and it was anticipated that the Federal Procurement Regulations with the "Buy America" provisions would have a cost impact on the fabricated steel components of this bridge. The regulations permit the acceptance of foreign steel if the cost of the domestic steel is more than twenty-five percent higher. The SAS contract allowed bids based on foreign material and fabrication, but only if a domestic bid is submitted as well. Again only one bid was received (from a joint venture headed by American Bridge) and even though the foreign bid met the differential cost requirement, it was far in excess of the budget.

Although the budget overrun on the acceptable bid on the SAS is only about five percent of the projected total for the entire East Crossing, given the current financial status of the state with its large debt, the SAS became the "whipping boy" for the budget overruns on all the toll bridge projects.

Several review and audit panels were appointed to decide what to do about the SAS. The consensus of these panels was that, while some saving might be achieved by redesign, the least risk in attaining the seismic safety objectives would be by rebidding the SAS with a number of specific revisions to the General Conditions: defederalize the bridge,[38] and revise the contract for such provisions as schedule, retention, damages for delay, and so on. In spite of these recommendations, Governor Arnold Schwarzenegger directed Caltrans to plan for an extension of the concrete Skyway to replace the SAS.

The East Crossing was designed as a global system to resist ground motion, with the SAS as a component within the system. The replacement of the SAS with a different structural component will introduce different boundary conditions at the interface with the Skyway and the Yerba Buena transition. In anticipation of this problem, Caltrans has cancelled work on Pier E2, the easternmost pier in the Skyway. The contract for the Yerba Buena transition has not yet been awarded, but Pier W2, the western-most pier in the transition, has been constructed under a separate contract.

The funding for the East Crossing was established by a Senate bill in the State Legislature. The bill that became law identified the SAS as an integral part of the East Crossing that was being funded. New legislation would have been required if the SAS was to be replaced. Although many in the Legislature could see the political and aesthetic advantages in retaining the SAS, there was disagreement as to how it should be funded. Finally, in July 2005, a compromise funding agreement was reached, and the Governor and the Legislature directed Caltrans to re-advertise the SAS. The current schedule is to advertise the contract in August with bid opening in November. Final completion of the entire East Crossing is scheduled in 2012.

---

38. Defederalization would mean that the bridge wouldn't use federal funds, hence wouldn't have to follow federal buy-American contracting rules, hence would save money on the steel.

NCEER and the Federal Highway Administration

Nicoletti: In 1997, the National Center for Earthquake Engineering Research (now called the Multidisciplinary Center for Earthquake Engineering Research), received a research grant from the Federal Highway Administration (FHWA) to look at criteria and guidelines for new construction, as well as existing construction. The work was essentially completed in 1999. Because it was an important project for NCEER, they set up a special review board, the Highway Research Council, comprised of an administrative group and a technical group. The Council met as a body several times, and the two groups met in executive session at each meeting. The technical group, which I chaired, had representatives from FHWA, Caltrans, academia, and from private practice. This NCEER project for FHWA was quite similar to the work done for Caltrans under ATC-32.

Scott: Did the work for the Federal Highway Administration encounter circumstances analogous to those noted in connection with NEHRP and the Building Seismic Safety Council, wherein there were strong pressures to water down the California-based standards and criteria? Should seismic standards that California and other western-area people think are pretty good be extended nationally into less seismically aware regions? Or should they be softened for such wider application? Did you encounter some of that with the federal highway work?

Nicoletti: There was some of that, but it wasn't quite the same because FHWA does not need a consensus group like the building codes do. They are an autonomous group and can set their own standards. The highway equivalent of a building code is the set of specifications issued by the American Association of State Highway and Transportation Officials (AASHTO). These specifications are adopted, in whole or in parts, by the Departments of Transportation of each state. The specifications must meet certain minimum FHWA standards in order to qualify for federal funds. In the past, however, Caltrans has followed its own seismic standards because they believe that the California standards are more stringent than the national standards.

NCHRP Project 12-49

Nicoletti: Recently, in 1999, AASHTO, in cooperation with FHWA, initiated the National Cooperative Highway Research Program (NCHRP) to be administered by the Transportation Research Board of the National Research Council. NCHRP has contracted with a joint venture of the Multidisciplinary Center for Earthquake Engineering Research (MCEER) and ATC to develop a comprehensive specification (NCHRP 12-49) for the seismic design of bridges. The new specification will be presented in the Load and Resistance Factor Design (LRFD) approach adopted by most current building codes. Ian Friedland was the project manager for the ATC/MCEER joint venture and I am a member of the project engineering panel, which consists of practicing engineers, academics, and representatives of state highway agencies.

The project is interesting in that it is attempting to incorporate many of the ground motion and analytical provisions developed in the recent FEMA documents, as well as the provisions proposed for Caltrans in ATC-32. Caltrans, historically, has adopted the AASHTO specifications for bridge design, but implemented their own seismic provisions rather than those of AASHTO. NCHRP 12-49 recognizes this but believes that many of the California seismic provisions are too restrictive to be specified in the national code. Their solution is to exclude California from the seismic provisions. Those of us from California on the project engineering panel do not agree with this and believe that AASHTO should acknowledge the Caltrans provisions and modify them for the rest of the U.S.

The second draft of the revised specification was completed in September 2000, and the final draft is due in mid-2001. The provisions will go through several internal AASHTO reviews and will probably be published as an appendix for information prior to being adopted.

Chapter 11

Some Engineering Personalities

...I have had the opportunity to meet and exchange ideas with many engineering professors from universities in various parts of the country.

Nicoletti: This may be a good time to mention some of the personalities that I have encountered in the course of my work in structural engineering.

Scott: Yes. That would be very appropriate at this point.

Nicoletti: In the course of my participation on a number of various national committees and boards, I have had the opportunity to meet and exchange ideas with many engineering professors from universities in various parts of the country. Among those that I have particularly enjoyed are the late Peter Gergely from Cornell; Mete Sozen, originally from Illinois, now in Indiana at Purdue; Jim Jirsa at the University of Texas; José Roesset from Texas A & M; Bob Hanson, originally from Michigan, now with FEMA; Larry Reaveley from Utah; Les Youd from Brigham Young; Helmut Krawinkler from Stanford; Freider Seible and Nigel Priestley from U.C. San Diego; I.M. Idriss from U.C. Davis; and Vitelmo Bertero, Egor

Popov, Bruce Bolt, Ben Gerwick, Joe Penzien, Jack Moehle, and Steve Mahin from U.C. Berkeley. From my experience in conducting workshops, I have a lot of admiration for anyone who teaches engineering. You don't realize how little you really know about a subject until you try to teach it to someone else.

For many years, in my capacity as Senior Project Engineer or Chief Engineer with the Blume firm, I was directly involved with many architects in business development, negotiations, and conceptual development of projects. Largely because of John Blume's reputation in earthquake engineering, I had the opportunity to work with many well-known architects, including John Carl Warnecke, Welton Becket, John Portman, John Graham, and Edward Durell Stone. I found that most of them didn't understand or appreciate structural engineering, but they were concerned about what an earthquake could do to their buildings. I would like to say a few words about two architects that did understand and appreciate what we could do for them.

Two Architects Who Understood

Nicoletti: When we first met, Gerald McCue was the junior partner of Joe Milano, an architect who had worked with John Blume at the Standard Oil Company. Standard Oil had contracted with Milano and our firm to design a research facility in what later became the Chevron research complex in Richmond, California. Milano died shortly after the inception of the project and John was instrumental in convincing the client that McCue could complete the project. McCue and his subsequent firm, MBT Associates, eventually earned national acclaim for the design of state-of-the-art research facilities for such clients as Dow Chemical, Alza, Syntex, Stauffer, as well as Chevron Research, and we provided the civil and structural engineering services for most of their projects. McCue was appointed as Chair of the Department of Architecture at the University of California in Berkeley and later was the Dean of the Graduate School of Design at Harvard University.

Edward Maher, known as Ned to his many friends, was a graduate of the Naval Academy and the Beaux Arts School in Paris. Ned had many close friends in the military and was selected for many of their projects in the U.S. and overseas following World War II. We associated with his firm (Blanchard & Maher and later Maher & Martens) on a number of projects. In spite of his classical background, Ned was a very practical architect. He was very honest and straightforward in his architectural design, as well as his business dealings. I enjoyed working with him on a number of projects in the 1960s and until he passed away in the mid-70s. He always had many interesting stories and also was very fond of martinis, good cigars, and good food.

John Blume and Henry Degenkolb

Nicoletti: First I will mention John Blume, who certainly was the dominant personality I had contact with in my years with the firm. Also there were other people, like Henry Degenkolb, who was certainly a major figure in San Francisco. My impression of Henry was that, while at times we were at odds, as he himself said, "We may not agree on the method, but we seem to end up with the same type of building." I think Henry and John tended to approach things from different directions, but probably usually ended up at the same place.

John was probably more scientifically oriented, while Henry was practical and detail oriented, but I think the end results were very similar. I know that Henry opposed some of the things that John came up with, such as ductile concrete, dynamic analyses, and other innovations. But eventually, Henry adopted them and became an exponent of them.

Scott: Yes, there was a big hassle about ductile concrete after Blume published a book jointly with two other authors,[39] and then Degenkolb and Roy Johnston did a critique of it at the request of the steel people.[40]

You also mentioned dynamic analysis. Was Degenkolb reluctant to accept that at first?

Nicoletti: In the early days, Henry did not believe that dynamic analysis was necessary or should be done at all. He believed that a building could be designed with equivalent lateral forces and should be detailed to stay together—to be tied well so that everything maintains integrity during an earthquake. Those detailing provisions are good, and we believe that, too. I think Henry eventually came around and became one of the leading exponents of dynamic analysis, and certainly his firm is today.

Scott: Why do you think Degenkolb changed his mind?

Nicoletti: I think the San Fernando earthquake, with its multitude of building response records, brought out a lot of the things that John Blume had been preaching. Other earthquakes also brought out some of these things. There was reconciliation between some of the analytical approaches and actual experience in earthquakes. Also, research—as more and more people in the universities became involved with earthquake provisions, their approach was more and more like John's.

Scott: And dynamic analysis became more sophisticated and more practical?

Nicoletti: Yes. The advent of computers made dynamic analysis more practical. Also, as computers developed more and got faster and cheaper, they came into more widespread use. It became possible not only to do linear dynamic analysis, but also nonlinear analyses. All these things pushed the profession more and more into analytical procedures as opposed to prescriptive detailing procedures.

Prior to the San Fernando, California earthquake of 1971, Los Angles passed an ordinance requiring seismic instrumentation of all new buildings over a certain minimum height. As a consequence, instruments were in place that generated over 200 records of the 1971 earthquake. The National Oceanic and Atmospheric Administration (NOAA) and EERI jointly sponsored detailed studies of the earthquake's effects. Our office was selected to report on five instrumented buildings. We compared the ana-

39. John A. Blume, Nathan Newmark, and Leo H. Corning, *Design of Multistory Concrete Buildings for Earthquake Motions*. Portland Cement Association, 1961.

40. Henry J. Degenkolb and Roy G. Johnston, "Engineering Evaluation of the Portland Cement Association's Book: *Design of Multistory Reinforced Concrete Buildings for Earthquake Motions*." Prepared for the American Iron and Steel Institute, June 1963.

lytically predicted response with the actual recorded response and obtained very good correlation between the two.

Scott: These records provided an opportunity to confirm the analytical procedures as well as the code provisions?

Nicoletti: Yes. Three of the buildings we investigated were reinforced concrete designed with the new ductile provisions, and all three buildings performed well. One of the buildings, the Holiday Inn on Marengo Boulevard in Pasadena, California had a recorded displacement of the roof that was nine times the calculated displacement at which initial yielding of the reinforcement would occur, with no apparent structural distress. This kind of evidence made believers of engineers like Henry Degenkolb, who had been skeptical about both dynamic analysis and ductile concrete.

Scott: As a sidelight on Henry Degenkolb and his skepticism, he was a bit reluctant or cautious about putting a lot of faith in base isolation. He seemed to accept the theory behind it, but was dubious about staking too much on its performance until we had seen some base-isolated buildings go through actual earthquakes.

Nicoletti: I agreed with Henry, but I think the provisions being used for base isolation also recognize some of those fears and have built in a lot of conservatism. A big concern about base isolation is whether you can properly represent the maximum displacement that the building will experience. Base isolation essentially trades displacement for force. You change the period of the structure to a longer period so that you get larger displacements but smaller forces. The isolator, as well as the space around the building, has to be able to accommodate the displacements. That is one of the big concerns. The current provisions for base isolation include quite a bit of conservatism. The design earthquakes being used have a lot more conservatism in them than those being used in the code for design. They are using longer return intervals, such as 1,000-year return intervals as opposed to 475-year intervals.

I always enjoyed Henry, who was a very personable guy. He was very friendly, I think much more so than John, at least in the minds of those who did not really know John.

Scott: Blume could be a little distant or reserved with people he did not know well?

Nicoletti: Yes. I always enjoyed my relations with Henry, and had contact with him on committees, at conventions, and so forth. We got along very well. I never had any disagreement or any other problem with Henry. I have a lot of respect for Henry and the things he accomplished.

John Rinne

Nicoletti: John Rinne is another engineer I had contact with. I enjoyed my relations with him, and was with him on several committees, including BCDC. John Rinne was very much like John Blume—he was analytically oriented. I understand that he had a large part to play in the original version of the Blue Book.

Scott: Yes, and prior to the Blue Book, he also led the northern California group that drafted Separate 66.[41]

Nicoletti: At the time, I was aware that the work was going on because I was working for

John Blume then. I wasn't a part of the drafting process, but I remember helping John with some examples that were being tried out. We were trying out some of the new code provisions on the Alexander Building, and I ran some of those calculations for John.

My main contact with John Rinne was when he was chief engineer at Standard Oil. We were doing work primarily for the Chevron facilities in Richmond, and he was reviewing and approving our work. He was very thorough and careful, and I thought he had an open mind.

Pete Kellam

Nicoletti: One of the people with whom I had a good deal of contact was Pete Kellam, who was a well-known figure in SEAONC. He was more in my generation, and we had very close contact. We were on the SEAONC board together, were on many committees, and both were president of SEAONC.

Scott: I never knew much about him, but his name keeps popping up.

Nicoletti: Pete worked with Mike Pregnoff for many years, and later was a partner in the firm of Graham and Kellam. He was hard working, and the sort of fellow you would put on a committee if you wanted to make sure something was going to get done. He could get others to do their jobs, and in a very nice way. He was not a hard taskmaster, but was able to get work out of people. He was a very prominent figure in SEAONC, both in its administration and in the committees.

Nick Forell

Nicoletti: My first contact with Nick was about twenty-five years ago. One of our principal architectural clients, MBT Associates, was negotiating a contract for a new building at the Lawrence Livermore Laboratory and was told that our structural engineering fee was unacceptable. I told them that perhaps a smaller firm might be able to do the work for the reduced fee and recommended Forell and Elsesser, because Eric had worked in our firm for a number of years. I got a telephone call from Nick, who couldn't believe what I had done. A few years later, he and I participated in two earthquake reconnaissance trips to Mexico for the Guerrero and Oaxaca[42] and, after the Loma Prieta earthquakes, we were both involved with Caltrans in peer reviews and on the Caltrans Seismic Advisory Board. I had mixed feelings when EERI told me that Nick had asked that I conduct the interview for his oral history.[43] I was pleased that he wanted me to do it, but I was aware that he was in the terminal stages of pancreatic cancer and I thought it would be very difficult for both of us. As it turned out, I think we both enjoyed reminisc-

41. Anderson, Arthur W., John A. Blume, Henry J. Degenkolb, Harold B. Hammill, Edward M. Knapik, Henry L. Marchand, Henry C. Powers, John E. Rinne, George A. Sedgwick, and Harold O. Sjoberg, "Lateral Forces of Earthquake and Wind, *Proceedings, American Society of Civil Engineers*. Vol. 77, Separate No. 66, April 1951.

42. Nicholas Forell and Joseph Nicoletti, *Mexico Earthquakes: Oaxaca, November 29, 1978; and Guerrero, March 14, 1979*. Earthquake Engineering Research Institute, 1980.

43. *Connections, The EERI Oral History Series: Nicholas Forell*, Joseph P. Nicoletti interviewer. Earthquake Engineering Research Institute, 2000.

ing about similar or shared experiences. I enjoyed Nick very much; he could seem cynical and incisive, but in reality he had a great deal of empathy and concern for others. I think his daughter put it very well at his memorial service when she described him as "a teddy bear with an attitude."

Ed Keith

Nicoletti: When Ed Keith came to work for us in the mid-1960s, he had just been discharged from the army at the end of the Korean War, his mother had passed away, his father had abandoned the family, and Ed was working full time, supporting his younger brother, and completing his master's degree at Berkeley. Five years later, he and his friend from Berkeley, Bob Feibush, with a total of $3,000 between them, started a company that eventually became known as Impel. Fifteen years later, they sold the company to Combustion Engineering for $125 million. Ed was able to capitalize on the engineering expertise for nuclear power plants that he had helped to develop in our office when the demand for these plants was at its peak, but I believe that, if it wasn't nuclear power plants, he and Bob probably would have found something else and been just as successful.

Michael Pregnoff and Albert Paquette

Nicoletti: Mike Pregnoff is someone else in SEAONC whom I wish I had gotten to know better. I think he had a large influence and was looked at as sort of the senior statesman with respect to the committee work.

Al Paquette was another of the older engineers I had a lot of respect for. Again, he was a senior statesman. He would come to committee meetings, and agree or disagree with some of the things we were doing. We found out that we should listen to him.

Scott: In your comments on their senior statesman role, I take it both Mike Pregnoff and Al Paquette, when they were considerably older but still participating in committee meetings, were valuable critics of what was being discussed?

Nicoletti: Yes. We did not always agree with them, but we found that we had better listen, and if we did not agree, we had better be right! Yes. Their views were based on a lot of experience, and they were very serious about the rationale behind their design.

Scott: They knew why they were doing the things they did in their practice. If you wanted to argue for doing something else, you had better have your position well thought-out?

Nicoletti: I think that is a good way to put it.

Nathan Newmark and George Housner

Nicoletti: I have also been fortunate enough to be associated on various projects with Nate Newmark and George Housner. I found both of them to be real gentlemen, and very modest about their significant accomplishments as pioneers in earthquake engineering. I first met George in 1972 when we were doing work at the Savannah River Nuclear Power Plant for DuPont. George had developed the original seismic criteria and DuPont asked us to meet with him to discuss our approach to the proposed retrofit work that we were doing. Later, as I have mentioned, he chaired the Caltrans Seismic Advisory Board and I got to know him better. George, like John Blume, is very much interested in the analytical derivation of seismic response, but also in finding a practical and rational approach to represent the response.

Chapter 12

Personal Reflections

I believe that today's job applicant, with a master's degree from a reputable university, has a far better foundation in structural engineering than we had.

Nicoletti: At the beginning of the project where we were evaluating naval housing in Europe, I was working almost full time at the San Francisco office as the only one assigned to the project, but as the evaluation procedure developed and additional personnel could be assigned, I began to think about retirement and gradually put the project into the capable hands of Ken Honda. Over the next year or so I was able to significantly reduce my time at the office until my retirement in July of 2003. During these last few years with the URS/Blume group I was able to devote a good deal of my time to advising and assisting the younger engineers and I find that I miss that relationship much more than the technical work.

Since my retirement I have been moderately active in committee work for ATC, BSSC, and MCEER, as well as the Caltrans Seismic Advisory Board and the Bay Bridge Peer Review Panel for Caltrans. I also am enjoying the additional time I now have for my family and the endless task of trying to keep my garden from turning into a jungle.

Scott: Tell us a little about your family and your nonprofessional interests and activities.

Nicoletti: As I mentioned earlier, my wife and I were married in Coeur d'Alene, Idaho at the end of World War II, and we will be celebrating our fifty-sixth anniversary this year (2001). My wife, Josephine, was a Navy nurse when we met during World War II, but after we were married, she didn't get a chance to practice nursing again until the kids were grown and away from home. She then did volunteer work at San Mateo General Hospital and conducted utilization reviews for MediCal until she decided to retire in 1975. We have three surviving children; we lost our son, Peter, in an auto accident in 1965. Our older daughter, Stephanie, is a nurse midwife practicing in Portland, Oregon. She has three daughters; the oldest has just started her own medical practice in New Brunswick in eastern Canada and last year presented us with our first great grandson. Our son, Dave, and his wife are high school teachers in Flagstaff, Arizona, and they have a little girl just entering the fourth grade. Our youngest, Mary Jo, last year married a technician in the Air Force and is working in the Planning Department of the City of Henderson, Nevada.

Scott: How about your own hobbies and interests?

Nicoletti: I have always been interested in most sports. I ran track and played basketball in high school, but in college I confined myself to intramural sports, although I was one of the "gym rats" looking for a pickup game of basketball whenever I could find time after class. After severing my Achilles tendon in an intra-office basketball game in 1967, I had to give up basketball, and a few years ago, I gave up volleyball. I also enjoyed downhill skiing until a few years ago when my legs told me that I had to give that up also.

I enjoy gardening; I worked in a wholesale flower nursery when I was in high school, and I like to think that I know more about flowers than I actually do, so I spend quite a bit of time taking care of my garden.

My wife and I both enjoy traveling and we have traveled extensively in Europe. In my first trip back to Italy in 1968, I found that I could understand the language, but I had trouble expressing myself. I enrolled in a couple of conversational Italian classes at our local community college and it helped quite a bit. I also took similar classes in German and Spanish to refresh what I had taken in school, and more recently I also took a French class. I enjoy other languages, and my superficial knowledge of Spanish has been helpful in the projects we have had in South America and the earthquake reconnaissance trips I made to Mexico and Nicaragua.

I have been involved with code development for the past thirty years with the SEAOC Seismology Committee, ATC, FEMA, BSSC, and ASCE. During this period, I have seen more and more involvement by the academic community in the various code committees. The result has been the introduction of many complicated provisions as we were provided with more and more analytical and research data by the academics. Many engineers believe that the codes have become too complicated, and would like to go back to the simpler provisions that may not have been rigorously correct, but were easier to understand.

In my personal experience, life was indeed easier when our office was able to design a building to the simplistic code provisions and then require that it conform to our own criteria, which were later published in the Department of Defense manuals discussed earlier. Codes have evolved from the representation of minimal seismic provisions to almost rigorous analyses of component responses. The older engineers believe that they are no longer able to include judgment in their designs because they no longer understand the basis of many of the provisions.

I personally think that this is a very exciting time in structural engineering as we are literally in a revolution with respect to seismic provisions. On one hand, I have to endorse many of the new provisions because they represent procedures that I have been doing for many years as a supplement to the old simplistic code requirements, but on the other hand, I believe that, for general application, we have to find ways to make them more transparent and easier to implement.

In the more than fifty years during which I have been practicing structural engineering in San Francisco, I have been able to observe many young engineers begin their careers in our office. Those that were my contemporaries in the 1940s and '50s usually started with only a B.S. degree, but their curriculum included a number of practical design courses and, armed with their slide rule and knowledge of moment distribution, they were ready to go to work. In the days before field welding and Simpson ties, the proper joinery and detailing of steel or timber connections was not only an art, but also a very necessary prerequisite of design. In fact, many engineers would start with the connection details before the design and sizing of the members.

Today's engineering graduate, with a B.S. and M.S. degree, is well versed in structural dynamics, matrix algebra, nonlinear analysis, and the user's manuals for SAP 2000 and ETABS, but lacks any practical experience in structural design or connection detailing.

However, I believe that today's job applicant, with a master's degree from a reputable university, has a far better foundation in structural engineering than we had. That is particularly true of earthquake engineering, which was practically unknown fifty years ago.

Photographs

The village of Ponte San Pietro, Italy, where Joe Nicoletti was born in 1921. The village straddles the old Roman road leading north from the walled city of Lucca and derives its name from its location by the bridge over the Serchio River on the outskirts of the city.

Joe Nicoletti upon graduation from grammar school, Daly City, California.

In the Boy Scouts.

On the University of California at Berkeley campus, at Sather Gate.

As a student at San Mateo Junior College.

Nicoletti with a dugout boat (banca) he and a friend purchased in the Philippines in World War II.

A lieutenant (jg) in the U.S. Navy in World War II.

Nicoletti, as the First Lieutenant of the USS Oglethorpe.

Josephine and Joseph Nicoletti in 1945 in Staten Island, New York, after their recent marriage in Coeur D'Alene, Idaho.

Relaxing on the beach, circa 1960.

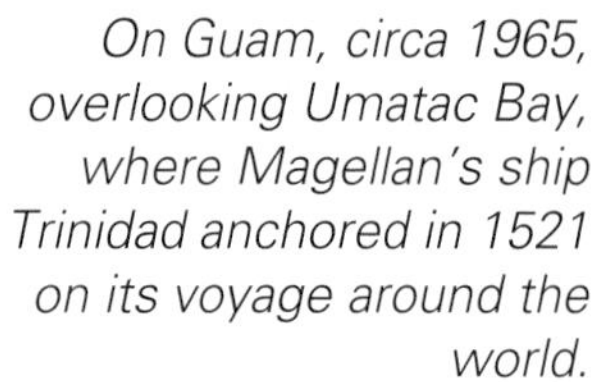

On Guam, circa 1965, overlooking Umatac Bay, where Magellan's ship Trinidad anchored in 1521 on its voyage around the world.

On Tinian Island in the Marianas, on the way to work on an engineering project in Saipan.

In his John Blume and Associates office in the Palace Hotel in San Francisco, circa 1975.

Architectural model of the Hyatt Regency Hotel in San Francisco.

Architectural model of the Federal Office Building in San Francisco.

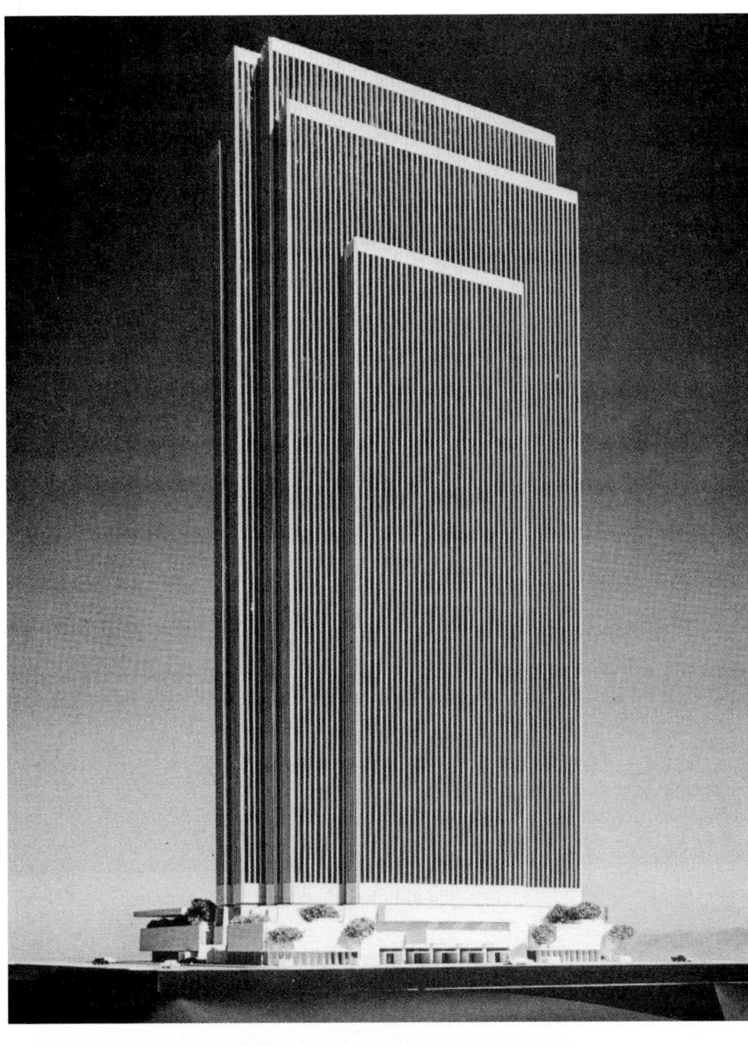

Architectural model of Embarcadero Four.

Typical installation of eccentric bracing in Embarcadero Four.

Above: Joe Nicoletti and Nick Forell in 1978 in Oaxaca, Mexico. Joe and Nick were co-authors of the EERI reconnaissance report on the 1978 Oaxaca and 1980 Guerrero earthquakes.

Left: Joe Nicoletti, left, Leonardo Zeevaert, and Nick Forell in Mexico City, circa 1978. Zeevaert was the lead structural engineer for the Torre Latinoamericana in Mexico City.

Josephine and Joe Nicoletti.

Appendix

Table 1

Applied Technology Council Presidents

Stephen E. Johnston	1973	Robert F. Preece	1990
William W. Moore	1974	Arthur E. Ross	1991
Joseph Kallaby	1975	Thomas G. Atkinson	1992
John F. Meehan	1976	Nicholas F. Forell	1993
Walter D. Saunders	1977	John M. Coil	1994
Joseph P. Nicoletti	1978	Edwin T. Huston	1995
Richard Christopherson	1979	John C. Theiss	1996
Daniel Shapiro	1980	C. Mark Saunders	1997
Bruce C. Olsen	1981	Charles H. Thornton	1998
Samuel Schultz	1982	Edwin T. Dean	1999
Howard Simpson	1983	Arthur N. L. Chiu	2000
Ajit S. Virdee	1984	Andrew T. Merovich	2001
James A. Willis	1985	James R. Cagley	2002
William T. Holmes	1986	Stephen H. Pelham	2003
T. Robert Kealey	1987	James M. Delahay	2004
Philip J. Richter	1988	Lawrence G. Griffis	2005
Lawrence D. Reaveley	1989		

Applied Technology Council Executive Directors

Roland Sharpe	1973-1979	Christopher Rojahn	1981-present
Ronald Mayes	1979-1981		

Table 2

Selected "alumni" (former employees) of the John Blume firm who went on to careers elsewhere

Jim Clark, Clark Pacific Precast Concrete
Jim Cooper, Federal Highway Administration
Raj Desai, Raj Desai Associates
Eric Elsesser, Forell-Elsesser Engineers
Sig Freeman, Wiss Janney Elstner
Ron Gallagher, R.P. Gallagher Associates
Gary Hart, UCLA, Hart Associates
Larry Kahn, University of Michigan
Anne Kiremidjian, Stanford University
Rich Klingner, University of Texas at Austin
Onder Kustu, Applied Technology Council, Oak Engineering
Gus Lee, Author, *China Boy*
Shi Chi Liu, National Science Foundation
Fritz Mathiessen, UCLA and USGS
Jack Meehan, California Office of the State Architect
Andrew Merovich, Merovich and Associates
Norm Owens, San Francisco State University
Jon Raggett, University of Santa Clara
Bruce Redpath, Geophysicist
Les Robertson, Skilling Helle Christiansen & Robertson
Roger Scholl, Applied Technology Council, CounterQuake
Roland Sharpe, EDAC, Applied Technology Council
William Spiller, New Jersey Institute of Technology
John Wiggins, J. H. Wiggins Co.
Fred Willsea, Fred Willsea Structural Engineer
Peter Yanev, EQE
Ted Zsutty, San Jose State University

Numerics

A

B

D

E

F

G

H

I

J

K

L

M

N

S

T

U

V

W

Y

Z